QUANTUM
ENTANGLEMENT

QUANTUM
ENTANGLEMENT

JED BRODY

The MIT Press | Cambridge, Massachusetts | London, England

This book was set in Chaparral Pro by Toppan Best-set Premedia Limited. Printed and bound in the United States of America.

Library of Congress Cataloging-in-Publication Data

Names: Brody, Jed, author.
Title: Quantum entanglement / Jed Brody.
Description: Cambridge, Massachusetts : The MIT Press, [2020] | Series: The MIT press essential knowledge series | Includes bibliographical references and index.
Identifiers: LCCN 2019024770 | ISBN 9780262538442 (paperback) | ISBN 9780262357616 (ebook)
Subjects: LCSH: Quantum entanglement.
Classification: LCC QC174.17.E58 B76 2020 | DDC 539.7/25--dc23
LC record available at https://lccn.loc.gov/2019024770

10   9   8   7   6   5   4   3   2   1

CONTENTS

SERIES FOREWORD

The MIT Press Essential Knowledge series offers accessible, concise, beautifully produced pocket-sized books on topics of current interest. Written by leading thinkers, the books in this series deliver expert overviews of subjects that range from the cultural and the historical to the scientific and the technical.

In today's era of instant information gratification, we have ready access to opinions, rationalizations, and superficial descriptions. Much harder to come by is the foundational knowledge that informs a principled understanding of the world. Essential Knowledge books fill that need. Synthesizing specialized subject matter for nonspecialists and engaging critical topics through fundamentals, each of these compact volumes offers readers a point of access to complex ideas.

Bruce Tidor
Professor of Biological Engineering and Computer Science
Massachusetts Institute of Technology

PREFACE

I read *The Tao of Physics* in high school, and it left me hungry to understand the mathematical rigor that inspired mystical statements about quantum physics. I was equally unsatisfied in college physics courses, which had plenty of mathematical rigor but no mystical statements whatsoever. I wrote a term paper about quantum entanglement, which is mysterious if not quite mystical. While working on the term paper, I read classic articles about entanglement, but the information sank into my mind no further than the level that handles paraphrasing.

One reason I didn't understand quantum entanglement is that I had never done an experiment with entangled particles. Indeed, the laboratory is the place where abstract concepts crystallize into palpable significance between your hands. The laboratory is where nature answers the questions posed by theorists. It's impractical, however, for every interested person to do every interesting experiment. Now that I've done experiments with entangled particles, I hope I'm able to explain the phenomenon to anyone who's curious.

Physics lab instructors sometimes feel the need to justify their existence—our existence. We insist that laboratory education is illuminating in ways that can never be fully conveyed in the lecture hall. Gesturing wildly for emphasis, we lavish praise on instructional experiments. Lab teaches hands-on skills and proves that physics actually works. Direct experience with an experiment gets you to think about the underlying physics more than anything else does. Certainly, doing experiments with entangled particles intensified my fascination with quantum mysteries. And I never would have done the experiments without the assistance and encouragement of several organizations and individuals.

Teaching physics lab isn't nerdy enough, so every three years physics lab instructors come together to learn from one another, and generally geek out, at a conference organized by the Advanced Laboratory Physics Association (ALPhA). In summer 2012, the lab conference happened to be in Philadelphia when I was there anyway, visiting my parents. If the conference had been anywhere else, I probably wouldn't have gone. I guard my summer vacation as greedily as Gollum guards the ring. I decided to go to the conference only because it practically arrived at my door. It almost would have been more effort to *avoid* going.

I expected to feel a bit disgruntled about being at the conference. Although I genuinely enjoy physics, I enjoy vacation even more. To my amazement, the conference was as enjoyable as a vacation. The talks and workshops were illuminating and inspiring. At that conference, I learned about the entanglement experiments for instructional labs. I also learned that ALPhA sponsors three-day "immersions" to teach instructors how to set up the experiment. In summer 2015, I attended the immersion led by Professor Enrique Galvez at Colgate University. I think I learned more physics in those three days than in any other three-day interval in my life.

After completing an immersion, instructors are eligible to apply to the Jonathan F. Reichert Foundation for a grant to help purchase the lab equipment. I'm grateful for the assistance that the Reichert Foundation provided for my instructional lab. The Emory physics department, where I work, covered the rest of the cost.

The final catalyst for this book was Emory's Interdisciplinary Exploration and Scholarship (IDEAS) program, which organizes "sidecar" courses, each of them geared to explore a topic that intersects with (and is cosponsored by) two existing courses in different departments. I wanted to teach a sidecar course about quantum entanglement. I expected it to be cosponsored by my Advanced Lab course and a philosophy course. But I couldn't find any philosophy professors who were interested. Luckily, I found a willing collaborator in the English department: Dave Fisher, who was scheduled to teach Technical Writing. We created a sidecar course to examine the different ways people write about quantum entanglement. If I hadn't delved into the existing literature to prepare for the sidecar course, I never would have thought to write this book.

Erin Bonning, Michael Weissman, Alissa Bans, and Tom Bing generously read and commented on a draft of this book. I'm grateful for additional helpful discussions with colleagues, especially Sergei Urazhdin, Daniel Weissman, Keith Berland, Vincent Huynh, Luiz Santos, Ajit Srivastava, and Justin Burton. I take full responsibility, however, for any errors or imprecision.

Albert Einstein memorably described quantum entanglement as "spooky action at a distance."[1] I recall another memorable Einstein quote, this one from a letter he wrote to the philosopher Erik Gutkind in 1954. There Einstein explained his view that the scientist's "religious feeling takes the form of a rapturous amazement at the harmony of natural law, which reveals an intelligence of such superiority that, compared with it, all the systemic thinking and acting of human beings is an utterly insignificant reflection." In a book about physics, it may be irrelevant to observe that the rapturous harmony of the natural world is increasingly imperiled. And yet, I dedicate this book to the preservation and restoration of the natural world.

INTRODUCTION

Quantum physics describes the behavior (and misbehavior) of tiny things: atoms, photons, and electrons, to name a few. What electrons lack in size, they make up for in importance. Electrons are the glue in chemical bonds, so quantum physics is used to understand the chemical bonds that hold together metals, plastics, skin, and every other material. Electrons are the lifeblood of computer chips, for instance, so engineers use quantum physics to design faster, smaller devices. Wherever quantum physics is applied, it's unerringly accurate.

The most amazing feature of quantum physics isn't its accuracy or its usefulness, but its brazen defiance of our common sense. Quantum physics challenges our basic understanding of reality itself. And yet, quantum physics started off in a very mundane way, seeking explanations for dry, quantitative data.

For example, hydrogen gas can emit four colors of visible light: violet, blue, aqua, and red. Physicists had carefully measured the wavelengths of these four colors: 410 nanometers, 434 nanometers, 486 nanometers, and 656 nanometers. Surely there's a reason for these four specific numbers. But what is the reason? Physicists were scratching their heads. In 1885, a physicist even came up with

The most amazing feature of quantum physics isn't its accuracy or its usefulness, but its brazen defiance of our common sense. Quantum physics challenges our basic understanding of reality itself.

an equation that fit all four wavelengths, but there was no explanation for the equation. It was a purely empirical equation, with no theory behind it.

Finally, in 1913, Niels Bohr came up with a theory that explained the four wavelengths. He claimed that the electron in the hydrogen atom is constrained to have certain amounts of energy. The electron cannot gain or lose energy smoothly; it can only make "quantum leaps" from one allowed energy level to another. Whenever an electron drops from one energy level to a lower energy level, it releases the energy in the form of light. The light emitted in a single quantum leap is called a photon. A photon is the smallest possible quantity of light with a particular wavelength. More generally, the smallest possible quantity of something is called a quantum.

These new quantum ideas had already solved two other mysteries. Max Planck explained the wavelengths of light emitted by hot objects, and Albert Einstein explained how photons knock electrons off the surface of metals. But even as quantum physics accumulated triumphs and grew in sophistication, it began to hint at deep mysteries in the fundamental nature of reality.

The fundamental quantum equation, which was established by Erwin Schrödinger and then published in 1926, dealt with probabilities: the likelihood of an electron appearing one place or another. Probability was not unfamiliar; the outcomes of coin tosses are also given as probabilities. But once a coin lands, the side that faces up is an objective fact, regardless of whether anybody knows what it is. In contrast with this common understanding of objective facts, the new quantum theory began to hint at a fundamental unknowability or uncertainty in unobserved particles. This conundrum drove Schrödinger to complain about the implications of his own equation.

Schrödinger asks us to imagine a cat trapped in an opaque box with an "infernal machine." The machine includes a radioactive material that occasionally emits a particle that can be detected by a Geiger counter. If the Geiger counter detects a particle, it triggers the release of a poisonous gas, which kills the sacrificial cat. The radioactive emission is governed by quantum physics. Quantum theory can specify only the probability that a particle will be emitted to trigger the release of poison gas. But unlike a tossed coin, which lands heads or tails up regardless of whether anyone observes it, quantum predictions aren't so easy to interpret. Quantum theory implies that before a measurement is performed, somehow the particle is neither emitted nor not emitted, or (equivalently?) both emitted and not emitted. In this case, the poison gas is both released and not released, and the cat is both dead and alive. This confusing condition persists until a measurement is performed. But what constitutes a measurement?

The intervention of a conscious observer who looks in the box? Or simply the interaction of the emitted particle with the Geiger counter?

Making matters worse, in 1927 Werner Heisenberg showed that the more precisely an electron's position is known, the more uncertain its speed becomes. The electron seems committed to not being pinned down. When an electron takes the witness stand, it never agrees to tell the whole truth (both its position and its speed). But does its refusal to tell the whole truth hint at a deeper truth? Are quantum measurements like breezes through a curtain, giving us shifting glimpses of a reality that is never fully revealed?

Some scientists argue that quantum physics predicts outcomes of measurements and nothing more; we shouldn't even ask the question "What does it all mean?" At least, we shouldn't claim to know what particles are doing when we're not measuring them. This is a form of Bohr's "Copenhagen interpretation," though the Copenhagen interpretation itself has been interpreted different ways by different people.

People like Einstein were fed up with vagueness, uncertainty, and contradictions. If *1984* had already been written when these physicists were grappling with these qualities of quantum mechanics, Einstein would have accused his opponents of doublethink: "*Doublethink* means

Are quantum measurements like breezes through a curtain, giving us shifting glimpses of a reality that is never fully revealed?

the power of holding two contradictory beliefs in one's mind simultaneously, and accepting both of them."[1] Surely nature itself is not guilty of doublethink. Surely quantum physics can be massaged and refined, retaining its accuracy while eliminating the fuzziness and absurdity.

Einstein, uncharacteristically, was wrong.

THE NEGATIVE SPACE OF QUANTUM PHYSICS

The quantum contradiction of common sense takes many forms. An especially rigorous form occurs in experiments with entangled particles. Two particles are entangled if the measurement of one of them, for all practical purposes, instantly affects the other particle over any distance.[1] Einstein called it "spooky action at a distance." Even spookier: the measurements of the particles do not reveal properties that the particles had all along. Prior to measurement, the particles' properties are not merely unknown, they are undetermined; and the measurement somehow transforms them—the properties are no longer fuzzy but focused.

The purpose of this book is to empower you to deeply understand how our common-sense assumptions impose constraints—from which entangled particles burst free. In other words, this book explains what quantum physics is *not*. Our task is to paint the negative space of quantum physics, a space composed of seemingly plausible theories that cannot account for measured results. I'm using "negative space" the way an artist would, to indicate the space around a subject. Let's imagine a space full of concepts. If we draw a border around quantum physics, our everyday assumptions occupy the excluded space, the negative space. Surprisingly, irritatingly, or magically—depending on your disposition—our everyday assumptions are contradicted by experiments with entangled particles.

Mathematics is a vehicle through which our assumptions become experimentally testable. We need only logic and arithmetic to understand how our everyday assumptions are contradicted by measurements of entangled particles. This is a relief, and perhaps surprising, since harder math is required to understand rocketry, semiconductor devices, heat conduction, and many other topics. Unlike these technological topics, quantum entanglement addresses the fundamental nature of reality. Perhaps nature's apology for behaving so strangely at the deepest level is to make its negative space mathematically accessible to all of us.

Does the mathematics of quantum entanglement say something mystifying, or even mystical, about the universe? Or, rather, should we be mystified by the quantum contradiction of our everyday assumptions? To answer this question, we will dive deep into simple yet rigorous logic. We will see that our common-sense assumptions impose simple mathematical constraints on measurable quantities. These constraints are violated by both quantum theory and measured data.

Measurements of entangled particles contradict at least one of the following two assumptions:

1. Realism: Objects have properties that exist regardless of whether anyone is observing them. Observation merely reveals properties that the objects had all along.

2. Locality: The measurement of one object can't affect the measurement of another object that is arbitrarily far away.

The combination of the two is called *local realism*: the assumption that objects have definite properties, independent of our knowledge of them, and independent of measurements performed on other objects. Local realism is deeply embedded in our common sense. When I measure the length of my left foot, I determine the length it already had, without affecting the length of my right foot. And yet this common-sense claim would be exactly wrong if my feet were entangled particles. (Though my feet do become entangled in a different sense, whenever I attempt to dance.)

How can experiments contradict our everyday assumptions? In this book I intend to answer that question.

It's surprising that a philosophical assumption has mathematical consequences, which can be tested experimentally. But local realism isn't the only philosophical assumption with mathematical consequences. We might characterize geocentrism as a philosophical assumption:[2] "Everything must orbit our planet due to our own preeminence in the universe." It's not obvious that this assumption should have mathematical consequences. And yet, ancient and medieval astronomers labored mightily with the mathematical consequences. They had to explain why the other planets occasionally go into retrograde, backing up as if looking for something they dropped. The geocentric astronomers came up with hugely complex and surprisingly accurate mathematical models. Ultimately, however, the preponderance of evidence, and the preference for a simple unifying theory, forced astronomers to abandon the geocentric assumption. Similarly, as we'll see, experimental evidence forces us to abandon the everyday assumption of local realism.

This is not exceptionally light reading, but it is recreational, for the analogies to quantum systems take the form of logic puzzles. Unlike most logic puzzles, these quantum analogies are not arbitrary or contrived but symbolize the predictions of quantum physics and thus the ultimate nature of reality. Besides analogies, we will study actual laboratory observations, including the first experimental contradiction of local realism.

All physicists agree on the mathematical predictions of quantum mechanics (which is what physicists call quantum physics). And all physicists agree that experiment resoundingly confirms these quantum predictions. But there is no consensus about how to interpret quantum mechanics. By the end of this book, you will understand the reasoning that forces us to discard everyday assumptions, and you will be able to draw your own conclusions.

Before journeying into the microscopic world, let's think about common objects like computers and coins. In the context of these familiar objects, we'll develop the concepts that we'll need later. In subsequent chapters, we'll study particles that are directly governed by the unnerving and enchanting quantum laws.

Hidden Variables

Let's begin by imagining a computer that displays a random number every time we tap the space bar. The displayed number is only one digit, so there are ten possibilities: 0, 1, 2, 3, 4, 5, 6, 7, 8, and 9. Each of these numbers is just as likely as the rest, appearing about 10 percent of the time. Suppose that we're really intrigued by these random numbers, and we want to develop a theory to explain what's going on inside the computer.

Our first attempt at a theory can simply be a summary of our observations: each number from 0 through 9 occurs about 10 percent of the time. This theory is a probabilistic theory: it gives us the probabilities of future outcomes, but it doesn't predict exactly what each outcome will be. This is a disappointment. We suspect that if we knew the algorithm used by the computer program, we could predict the exact outcome every time we tapped the space bar.

So, we begin to speculate about possible algorithms employed by the computer program. Suppose the seemingly random numbers are based on a hidden clock inside the computer. The clock, we imagine, has a precision of milliseconds, so the time starts at 0.000 seconds and proceeds to 0.001 seconds, and so on. What if the seemingly random number is simply the last digit of the hidden clock at the instant we tap the space bar? So if the hidden clock has the value 143.852 when we tap the space bar, the computer displays the last digit, 2. If the hidden clock has the value 5762.267 when we tap the space bar, the computer displays 7. The displayed numbers appear random because the last digit of the hidden clock is completely uncorrelated with our casual decision to tap the space bar.

Let's further develop this theory, even though it's a just a guess. Let's say that the last digit of the hidden clock is a *hidden variable*. The theory that we're developing is a *hidden variables theory*. Let's call the hidden variable λ

(lambda), so when the hidden clock has the value 143.852, λ = 2. When the hidden clock has the value 5762.267, λ = 7. Let's say that the number displayed on the computer monitor is N. So our hidden variables theory is simply

$N = \lambda$:

the number displayed on the monitor equals the hidden variable.

In the hidden variables theory we just came up with, λ isn't so hidden because it happens to equal the displayed number N. We can imagine an alternative theory in which

$N = 9 - \lambda$

and λ is still the last digit in the hidden clock. But now, when the hidden variable λ = 0, the displayed number N = 9. Our new hidden variables theory generates seemingly random numbers just as well as the original theory. In fact, there are lots of other equations we could think of to explain the random number N in terms of the hidden variable λ. Here's just one more example: $N = \lambda+1$ unless λ = 9, in which case N = 0. There are many equally plausible hidden variable theories (equations for N in terms of λ).

We can even consider other sources of the hidden variable λ. Maybe λ, instead of being based on a hidden clock, is somehow based on the number of people watching internet videos of puppies at the moment we tap the space bar; the computer program could search online to find this information. Or λ could be based on the price of gold, the temperature in Nairobi, or countless other quantities found online or inside the computer.

We can imagine many quantities that λ could be based on, and we can imagine many ways to calculate N in terms of λ. In all cases, we remain confident that the computer program uses some algorithm to determine the displayed number N based on some hidden number λ; the computer doesn't pull N out of thin air.

We originally defined N as the number displayed on the monitor, but we've now redefined it as a quantity calculated from λ; λ changes over time, though at any moment λ has a precise value. Since N can always be calculated from λ, N too has a precise value at every moment, even though we're not tapping the space bar all the time. So N is a quantity that exists, at least mathematically, at every moment, even when we're not observing what it is.

Let's compare the hidden variables theory with our initial probabilistic theory, which simply stated that each number from 0 through 9 occurs about 10 percent of the time. The probabilistic theory is *true* but *incomplete*; it doesn't predict the displayed number with certainty, so the theory's missing some facts. The hidden variables theory is complete because it does predict the displayed number with certainty, at any moment. For example, we could propose the hidden variables theory—that the displayed number N is the last digit of the number of people currently watching internet videos of puppies—and then we could test this theory by finding out how many people were watching puppy videos at the moment we tapped the space bar. The hidden variables theory may be correct or incorrect, but it always predicts exactly what N is at every moment.

Let's suppose we come up with an accurate hidden variables theory. Then, the probabilistic theory is consistent with the hidden variables theory. Specifically, the hidden variables theory predicts N with certainty, and the probabilistic theory correctly gives the probabilities of the different possible values of N.

Flipping a Quarter

Imagine flipping a quarter. (But you can do more than just imagine: the equipment required for this experiment costs only 25 cents! If you can't afford a quarter, you can try selling this book for a quarter. If no one thinks it's worth a quarter, try a dime. A nickel? A penny? Anyone? Please?)

You flip the quarter into the air using your thumbnail. While the quarter is spinning in midair, you use one hand to slap it onto the back of your other hand. Don't remove the top hand yet; keep the quarter concealed. Without looking at the quarter, you already know that the outcome of the coin toss is either heads or tails. We'll say that these are the two possible *states* of the quarter.

Do you believe that the state of the quarter is already determined, even though you can't see it through your hand? Or is the state of the quarter not only unknown but also *unknowable* prior to observation? Is direct observation the only reality? Does the definite state of the quarter come into existence only when it's observed? Most people probably believe that the state of the quarter is fixed the moment you slap it between your hands. We don't know the state of the quarter until we look at it, but the observation merely reveals the state that the quarter is *already in*. If we believe that the concealed quarter has a definite state even though no one's observing it, then we believe in *realism*. According to realism, physical states exist totally independently of whether anyone is observing them (or whether any laboratory instrument is measuring them). If we don't know the state of the quarter hidden by our hand, it's only because we're ignorant; the quarter has a definite, physical state that we happen not to know.

Let's try to outline a theory to predict the quarter's state. Our first attempt might be to simply specify what we observe after repeated trials: about half the time we get heads, and the other half of the time, we get tails. This is a probabilistic theory, and we want to try to do better. We want our theory to conform to physical reality. If the

Is the state of the quarter already determined, even though you can't see it through your hand? Or is its state not only unknown but also *unknowable* prior to observation?

quarter *has* an exact physical state even before we observe it, then our theory should *predict* the exact physical state before we observe it. The probabilistic theory is accurate, but it's incomplete because it doesn't determine individual outcomes with certainty.

If we want to predict the quarter's state before observing it, we have to do a lot of physics. We have to consider all the forces on the quarter: the flick of our thumbnail, the weight of the quarter, possibly the effect of air resistance, and the force of the top hand as it comes down onto the quarter. We also have to know the position on the quarter at which each force acts: Do we flick the quarter right near the edge, or toward the center? We need to know the initial orientation of the quarter. There may be other parameters that we haven't even thought of. We don't actually want to go through all the work of developing a complete theory, so we don't need to list every parameter that influences the final state of the quarter. Instead, we refer collectively to all the influential parameters as hidden variables.

Without even constructing a hidden variables theory, we can say some things about it. It is complete: it predicts with certainty the final state of the quarter every time. It conforms to realism: it assigns a final state to the quarter regardless of whether anyone is observing it. It is consistent with the probabilistic theory: it predicts heads half the time, and tails half the time. In the hidden variables theory, though, nothing is truly random.[3] Instead, common variations of the hidden variables (the speed and angle at which the thumbnail flicks the quarter, etc.) lead to a prediction of heads just as often as tails.

Shaking Two Quarters

Now imagine shaking two quarters between your hands. (The cost of the experiment just doubled. At this rate, you'll need a particle collider by the end of the book.) You know that one of the quarters is dated 1999, and the other is 2000. You shake the quarters so well that you don't know which is which. Without looking at the quarters, you separate them, one in each hand, concealed by your fingers.

Before you open either hand to observe a quarter, is the date of that quarter already a physical reality? Or do the properties of the quarter come into existence only when you observe them? Once again, we're asking about realism: Is the date of the quarter in your left hand an objective, physical reality even if no one can see it or know with certainty what it is?

If we reject realism in this case, something very strange happens. If a specific property (the date) of the quarter in your left hand comes into existence at the moment it's observed, then the quarter in your right hand must simultaneously acquire the *other* date. The observation of one quarter, absurdly, affects the other quarter.

Common sense rejects absurdity and demands that the observation of one quarter has no effect on the distant quarter (even if the distance is only an arm's length). This is the everyday assumption of *locality*: the observation of one object has no effect on a distant object. In fact, locality implies realism. If observing one quarter has no effect on the other quarter, and the two quarters are always observed to have different dates, then both quarters must have had their dates all along. This combination of locality and realism is *local realism*: objects have properties that exist regardless of whether anyone's observing them, and they're unaffected by observations of distant objects.

Let's contrast a few different points of view. Suppose you shake the two quarters and conceal one in each hand. After a moment of quiet contemplation, you open your left hand to reveal the quarter dated 1999. What does this imply about the quarter in your right hand, which you still haven't opened?

• Local realism: The quarter in your left hand was the 1999 quarter all along, and the 2000 quarter was in your right hand all along. The observation of the 1999 quarter merely informs us where each quarter was all along. The observation has no effect whatsoever on either the quarter you observe, or on the other quarter.

• Observation transforms physical reality: The quarter in your left hand became the 1999 quarter at the

This combination of locality and realism is *local realism*: objects have properties that exist regardless of whether anyone's observing them, and they're unaffected by observations of distant objects.

moment you observed it. The quarter in your right hand simultaneously transformed into the 2000 quarter. We dare not ask what the dates on the quarters were prior to observation.

• Direct observation is the only reality: Unobserved properties don't exist; anything outside our senses is senseless, and is imagination, not reality. The quarter in your left hand acquired its property (the date, 1999) the moment you observed it. The unobserved quarter in your right hand still has no properties (beyond what you can feel), even though you know that whenever you observe it you'll see the date 2000.

It seems that all of these viewpoints, however implausible, cannot be refuted by evidence: How can we acquire any evidence of how something is before it's observed, given that we gather evidence only through observation? Astonishingly, astoundingly, physicists have discovered:

• Local realism imposes constraints on measurable quantities. (Many versions of this surprising fact are proven in detail later in the book.)

• Measurements of entangled particles violate the constraints imposed by local realism.

• Therefore, local realism is not a valid assumption for entangled particles.

This book will empower you to draw your own conclusions from the (almost) incontrovertible fact that quantum mechanics is incompatible with local realism.

Let's apply local realism to the two quarters. (Since quantum mechanics doesn't apply to quarters in an obvious way, local realism is a legitimate viewpoint.) If the 1999 quarter was in your left hand even before you peeked at either quarter, then some combination of forces drove that quarter into your left hand. In principle, we should be able to develop a theory of hidden variables to predict which quarter ends up in which hand. The hidden variables include the starting positions of the quarters before you shake them together, the exact shape of your cupped hands, and the rate and intensity of the shaking. We want to emphasize that the observation of the quarter in one hand does not affect the quarter in the other hand, so we call our theory a *local* hidden variables theory. The local hidden variables theory is consistent with the simple probabilistic theory that merely gives the 50 percent probability of each possible outcome. The local hidden variables theory goes beyond the probabilistic theory to predict the *exact* outcome of each trial.

Before proceeding to real particles and their entanglement, let's review the main points from this chapter.

• A hidden variables theory assumes that a seemingly random process is actually governed by influences that we're inadequately aware of. If we knew all the influences exactly, we'd be able to predict all outcomes exactly.

- An accurate hidden variables theory is consistent with the observed probabilities of outcomes. For example, a hidden variables theory describing a coin toss should predict heads half the time—as well as predicting exactly *which* tosses will be heads.

- Realism is the belief that objects have properties that exist totally independently of whether anyone is observing them (or whether any laboratory instrument is measuring them; I use "observe" and "measure" interchangeably).

- Locality is the belief that the observation of one object can't affect a distant object.

- If two quarters are shuffled and then separated, locality implies realism: if the observation of one quarter has no effect on the other, and the two quarters are always observed to have different dates, then they must have had their dates all along, prior to observation.

- As subsequent chapters will demonstrate in detail, the assumption of local realism imposes constraints on measurable quantities. These constraints are violated by measurements of entangled particles.

AN EXPERIMENT TO CHALLENGE A PHILOSOPHY

We know that electrons have an electrical property: negative charge. Electrons also have a magnetic property called spin. In the first experimental observation of spin, a beam of silver atoms was aimed through a magnetic field that was stronger in some places than others.[1] (We can create this kind of magnetic field by taking a horseshoe magnet and sharpening one of the poles.) If we send silver atoms between the poles of the magnet, we find that some atoms are deflected toward the north pole, and some are deflected toward the south pole (figure 1). No atoms pass straight through. Just like a coin toss, exactly two outcomes are possible.

This experiment is generally performed with neutral atoms—not isolated, negatively charged electrons. The deflection resulting from an electron's charge overwhelms the deflection resulting from its spin. But the beauty of

a. Deflection toward the north pole.　　b. Deflection toward the south pole.

Figure 1　The arrows represent silver atoms passing through a magnetic field, with some deflecting toward the north pole (a), and some deflecting toward the south pole (b).

thought experiments is that we can ignore inconvenient details such as these. Let's do a series of thought experiments in which the electron's deflection results from its spin, and we ignore the deflection caused by its charge.

We imagine an experiment with special pairs of electrons. Suppose that the two electrons are emitted from a common source, and they travel in different directions. Each electron encounters a magnetic field. If the two magnets have the same alignment, the two electrons are always deflected in opposite directions: one to the north, and one to the south (figure 2). Because the behavior of one electron is linked so powerfully to the behavior of the other, we say the electrons are entangled.

In subsequent chapters, we'll discuss experimental methods for creating pairs of entangled particles. In the laboratory, it's easier to do experiments with entangled

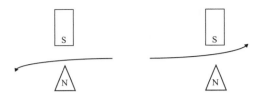

Figure 2 A pair of entangled electrons with opposite magnetic properties: if the magnets are aligned, one electron must be deflected to the north pole, and one must be deflected to the south pole.

photons than with entangled electrons. For now, however, we'll again set practical concerns aside.

Imagine observing many pairs of entangled electrons passing through the magnetic fields. Half of the time, the electron on the left is deflected north and the electron on the right is deflected south. The other half of the time, the opposite occurs: the electron on the left is deflected south and the electron on the right is deflected north.

Here's the key question: Does each electron have the magnetic property that is eventually observed (predisposition to northward or southward deflection) all along, or do the electrons "make up their minds" when they reach the magnets? In other words, does the measurement show us what the electrons were like all along, or does the measurement fundamentally change something? This may seem like a silly question, or at least one that cannot be empirically tested. What difference does it make whether the electrons were in the measured state all along, and what's the big deal if they're in an undecided state until the last minute? Indeed, I've been in an undecided state about what to order in a restaurant, and it's only the "measurement" taken by the server that forces me to make up my mind.

This is the big deal: If the electrons make up their minds at the last minute, they must make opposite decisions. If one chooses northward deflection, the other must choose southward. How can they coordinate this, in defiance of locality, when they're in different places? If the electrons make up their minds at the last minute and they always make opposite decisions, they're like twins with a telepathic link, if you'll forgive the analogy. This is "spooky action at a distance"—and Einstein argued strenuously against its existence.

To preserve locality (and avoid spooky action at a distance), we'd better hope for realism, which asserts that the electrons all along have the properties we end up measuring. In 1935, Einstein argued that quantum mechanics is an incomplete theory,[2] like the probabilistic theories about quarters we considered in chapter 1. His argument is that the electrons are deflected in opposite directions because they were created with opposite magnetic properties. Quantum mechanics can't tell us in advance which way each electron will be deflected; it can only tell us that each electron has a 50 percent chance of being deflected

To preserve locality (and avoid spooky action at a distance), we'd better hope for realism, which asserts that the electrons all along have the properties we end up measuring.

northward, and the other electron must be deflected southward (if the magnets are aligned).

A complete theory should predict the magnetic property of each electron at the outset, if we believe that each electron *has* a definite magnetic property at the outset. We don't know the details of this hypothetical, complete theory, and we don't even know what factors predetermine the properties of the electrons. Since the unknown factors are hidden variables, the hypothetical, complete theory is a hidden variables theory. More specifically, we are interested in a local hidden variables theory: a theory that predicts the outcome of the measurement of a single electron, without any dependence on the other electron in the pair. A local hidden variables theory is thus an expression of the assumption of local realism.

For decades, physicists assumed that a local hidden variables theory could, in principle, complement quantum physics, filling in missing information and replacing probabilities with certainties. But the issue seemed academic or philosophical, and not subject to experiment: a local hidden variables theory determines the state of an electron before you measure it. Is it possible to measure the state an electron is in, before it's measured? Seemingly it is not.

In 1964, John Bell made a stunning theoretical discovery, called Bell's theorem.[3] His original paper languished in obscurity for years, but enthusiasm for his discovery swelled over the course of decades. Bell showed that any local hidden variables theory imposes a constraint on measurable quantities. The constraint on measurable quantities is now called a Bell inequality. If the constraint is violated by measurement, then a local hidden variables theory cannot be valid. Moreover, because quantum physics predicts violations of the Bell inequality, quantum physics is fundamentally incompatible with any local hidden variables theory. So Einstein's hope was in vain: a local hidden variables theory cannot complement quantum mechanics; it can only contradict it. And since measurable quantities determine whether a Bell inequality is violated or not, an experiment can be performed to determine whether the real world is consistent with quantum mechanics, or with a local hidden variables theory; we can't have both. To reiterate these key points:

- A Bell inequality is a constraint on a measurable quantity. An experiment can be done to test a Bell inequality. The experiment will either satisfy the Bell inequality or violate the Bell inequality.

- If the experiment satisfies the Bell inequality, the experiment conforms to local realism but contradicts quantum physics.

- If the experiment violates the Bell inequality, the experiment may conform to quantum physics but must contradict local realism.

Experimental results indeed violate Bell inequalities, thereby confirming quantum mechanics and overruling any possible local hidden variables theory. What does this mean? Either locality fails, or realism fails, or both fail: either one electron is influenced by the distant electron or the distant magnet, or the act of measurement *creates* a definite magnetic property that the electrons did not previously have, or both of these strange phenomena occur. (This is not a comprehensive list of interpretations of quantum mechanics, but it's representative of possible consequences of rejecting local realism.)

Bell's original theorem is too mathematical to prove in this book. Luckily, physicists have developed simplified versions of Bell's theorem. We will see some of these in chapter 4, where we will prove that measured data contradict the philosophical assumption of local realism.

For now, we will investigate the fact that Bell established, without proving why it's true. Bell asks us to imagine rotating the magnets encountered by the electron pairs. So, for example, we could rotate one magnet 180° relative to the other. If we do this, we find that the two electrons are always deflected toward the same pole: both to the north, or both to the south (figure 3).

Now we can imagine rotating the magnets to any angle, not just 180°. Let's say a magnet is set to angle 0° if the south pole is directly above the north pole (like the magnets in figure 2). So the magnet is set to angle 180°

Either locality fails, or realism fails, or both fail.

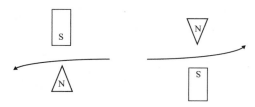

Figure 3 If one magnet is rotated 180° relative the other, the two electrons are always deflected toward the same pole: both north (as shown), or both south.

if the north pole is directly above the south pole. We can do experiments in which either magnet is set to any angle.

Bell asks us to think about just two numbers: +1 and −1. If the two electrons are deflected toward the same pole (both north or both south), we write down +1. If they're deflected toward opposite poles (one north and one south), we write down −1. So, if the two magnets are aligned, the electrons are always deflected toward opposite poles (figure 2), and we always write −1. If one magnet is flipped 180° relative to the other, the two electrons are always deflected toward the same pole (figure 3), and we always write +1.

Bell asks us to think about three different magnet angles. Let's choose 0°, 45°, and 90°. Then he asks us to do the following:

- Set one magnet to 0° and one to 90°. Watch a bunch of electron pairs as they pass through the magnets. For each pair, write down +1 or –1 to indicate electron pairs deflected toward the same poles or opposite poles, respectively. Then, average all these numbers. Call this average number A.[4] (Its value will be in the range of –1 to +1.)

- Set one magnet to 45° and one to 90°. Again watch electron pairs as they pass through the magnets, and write down +1 or –1 for each pair. Average all these numbers. Call this average number B.

- Set one magnet to 0° and one to 45°. Once again watch electron pairs as they pass through the magnets, and write down +1 or –1 for each pair. Average all these numbers. Call this average number C.

Now we have three numbers, A, B, and C, based on experimental measurements. Bell proved that the assumption of local realism requires

$-1 - C \leq A - B \leq 1 + C.$

That's it! That's the result of Bell's theorem, the original Bell inequality. So, we now have a method to test a philosophical assumption. We simply perform measurements and calculate the three average numbers A, B, and C, as instructed above. Then we put these numbers in the Bell inequality. If the result is true, then we've satisfied the Bell inequality, and the data are consistent with the assumption of local realism. But if our data violate the Bell inequality, then we've contradicted local realism.

Let's recall that local realism is an everyday assumption: observation merely reveals properties that an object already had, and the properties of an object are unaffected by the measurement of a distant object. This is the assumption that leads inexorably to the Bell inequality. The Bell inequality is, in fact, violated by measurement, so the assumption of local realism cannot be valid. We'll explore the mysterious implications later in the book.

Next, we will look at the entanglement of light. Entangled light was used in the first experimental tests of Bell inequalities.

ENTANGLED LIGHT

A crowd of unruly students seeks admission to a prestigious yet absurdly strict boarding school. This school rigorously upholds a tradition whose purpose is long forgotten: all students must carry a baton at all times, and they must hold it precisely upright. Each eager candidate in the crowd dutifully holds a baton, but these students, hopeful for admission, as yet lack discipline: only a few hold the batons upright. The scowling headmaster admits only half the students. The closer each student's baton is to vertical, the more likely the student is to get in. The batons of admitted students are nudged to the proper, vertical orientation, which the students assiduously maintain for the duration of their studies at the boarding school.

The time arrives for the students to apply to college. They seek admission to prestigious yet absurdly strict colleges that also require students to carry batons at all times, the tradition being as inviolable as it is arbitrary and meaningless. Unhappily, not all colleges require the batons to be carried vertically. Some require the batons to be held at an angle of 30° from the vertical, others require 45°, still others require 60°, and some even require horizontal batons. The students have trouble convincing the colleges that after long years of meticulously holding vertical batons at their boarding school they will be able to adjust to new angles. In fact, the likelihood of college admission declines dramatically as the required angular difference increases:

• A student from this boarding school is 100 percent likely to get into a college that requires batons to be held vertically.

• A student from this boarding school is 75 percent likely to get into a college that requires batons to be held at an angle of 30° from the vertical.

• A student from this boarding school is 50 percent likely to get into a college that requires batons to be held at an angle of 45° from the vertical.

• A student from this boarding school is only 25 percent likely to get into a college that requires batons to be held at an angle of 60° from the vertical.

- A student from this boarding school has no chance of getting into a college that requires batons to be held horizontally.

The batons are analogous to the electric field in a light wave. If all light in a light beam has the same electric field direction, the light is said to be *polarized*. If the electric field in the light wave always points in a vertical direction, that indicates vertical polarization. Most light sources emit unpolarized light, however, in which the electric field can point in any direction perpendicular to the light ray.

Now let's think about the particles of light called photons. In unpolarized light, the photons, like the unruly applicants to the boarding school, are polarized in all different directions. A polarizer, like the scowling headmaster, transmits only the photons able to comply with the required angle (figure 4); the remaining photons are absorbed or reflected. Polarizers can be made of a variety of materials, including sheets of plastic. An ideal polarizer transmits exactly half the photons in unpolarized light. All the transmitted photons are polarized identically, in the direction enforced by the polarizer.

Consider only the photons transmitted through a polarizer. If these polarized photons encounter a second polarizer, the transmission through the second polarizer is exactly analogous to the college acceptance of the baton-toting students:

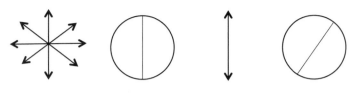

a. Unpolarized light   b. Vertical polarizer   c. Vertically polarized light   d. A second polarizer

Figure 4 (a): The electric field in unpolarized light can point in any direction perpendicular to the direction that the light is traveling. So here, imagine light is traveling into the diagram. (b): Any light passing through a vertical polarizer is (c) vertically polarized. (d): The fraction of vertically polarized light that passes through a second polarizer depends on the angle between the transmission directions of the two polarizers.

• A photon has a 100 percent probability of passing through a polarizer whose transmission direction is the same as the photon's polarization.

• A photon has a 75 percent probability of passing through a polarizer whose transmission direction is 30° different from the photon's polarization.

• A photon has a 50 percent probability of passing through a polarizer whose transmission direction is 45° different from the photon's polarization.

• A photon has a 25 percent probability of passing through a polarizer whose transmission direction is 60° different from the photon's polarization.

• A photon has a 0 percent probability of passing through a polarizer whose transmission direction is 90° different from (perpendicular to) the photon's polarization.

Quantum theory is consistent with these empirical facts. So if we like, we can think of these facts as quantum predictions.

Now that we understand polarization, we can understand measurements of entangled photons. To create pairs of entangled photons, we make use of a process in which a single photon splits into a pair of photons.[1] This splitting occurs when light travels through certain materials, such as crystals of *beta barium borate*. For example, a violet photon might split into two identical infrared photons, which travel away in different directions.[2] Only a small fraction of incoming violet photons split; most of them pass straight through.

We're going to be interested in the polarization of the pairs of infrared photons. Suppose we're shining a beam of polarized violet photons at a beta barium borate crystal. Let's polarize the violet photons 45° from the vertical. We observe that for a certain angle of the beta barium borate crystal, the infrared photons produced are vertically polarized (figure 5). If we rotate the crystal 90°, the infrared photons are horizontally polarized (figure 6).

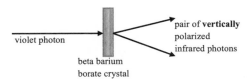

Figure 5   Some violet photons split in a beta barium borate crystal, producing a pair of infrared photons. For a particular orientation of the crystal, the infrared photons are both vertically polarized.

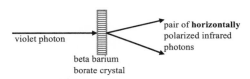

Figure 6   At a different angle of the crystal, the infrared photons produced are horizontally polarized.

Now we get two beta barium borate crystals and orient the first one to produce vertically polarized infrared pairs, and the second to produce horizontally polarized infrared pairs (figure 7). Each incoming violet photon has a possibility of splitting in the first crystal (producing a vertically polarized infrared pair), and an equal possibility of splitting in the second crystal (and producing a horizontally polarized infrared pair). Thus, the infrared pair is either vertically polarized or horizontally polarized.

How can we confirm this? We can put single-photon detectors in the path of the infrared photons. When a

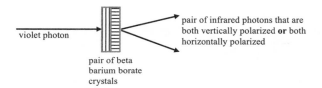

Figure 7 The violet photon may split in one of the two crystals, so the infrared photons produced may both be vertically polarized, or they may both be horizontally polarized.

photon reaches a detector, the detector produces a brief electronic pulse. If both detectors pulse at the same time, we call it a *coincidence* (which is the technical term for simultaneous detection at both detectors). It's very likely that a coincidence indicates a pair of infrared photons that split from the same violet photon.

Next, we put a polarizer in front of each detector (figure 8). If we make one polarizer horizontal and one vertical, we never see any coincidences. (This is an idealization, but let's keep things simple.) This confirms that no infrared pair consists of one vertically polarized photon and one horizontally polarized photon.

If we make both polarizers vertical, we see lots of coincidences, indicating vertically polarized pairs. If we make both polarizers horizontal, we see lots of coincidences, indicating horizontally polarized pairs. Let's say we see about 100 coincidences per second when both polarizers are vertical, and about 100 coincidences per second when

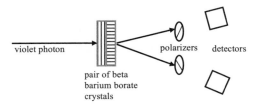

Figure 8 A polarizer is placed in the path of each infrared photon. If a photon passes through a polarizer, it reaches a detector. If the detector detects a photon, it sends an electronic signal to a circuit. If both photons are detected, they're detected at the same time, and a *coincidence* is recorded.

both polarizers are horizontal. When we remove both polarizers, we see about 200 coincidences per second: both the vertically polarized pairs and the horizontally polarized pairs.

All evidence indicates that we're creating pairs of photons such that both are horizontally polarized or both are vertically polarized. We want to know if the photons all along are predisposed to a particular measured outcome, or if the photons are in a fundamentally undetermined state prior to measurement.

Let's recall Einstein's argument in favor of realism and apply it in this context: Whenever both polarizers are horizontal, for example, the two photons in a pair must always do the same thing (pass through if they're horizontally polarized, or get blocked if they're vertically polarized); therefore, common sense dictates that the two photons all along must have shared properties that predetermine their behavior at the polarizers. If, instead, the photons decide at the last minute what their measurement outcomes will be, they appear to be in some mysterious collusion: spooky action at a distance.

Einstein insisted on realism to preserve locality: the photons must have shared properties all along—from the moment they are created in a single location. And since these shared properties can't be predicted with certainty in quantum mechanics, quantum mechanics must be incomplete, says Einstein: an unknown, more powerful theory should tell us exactly what each individual photon will do under all possible measurement conditions.[3]

Einstein tirelessly defended our common-sense assumptions, which were shared by many other prominent physicists. But we can no longer accept Einstein's argument in favor of local realism. Bell showed that local realism imposes constraints that may be either satisfied or violated by experiment. In fact, experiment violates these constraints (Bell inequalities), so local realism is overthrown in the laboratory. John Clauser and his PhD student Stuart Freedman achieved the first experimental violation of a Bell inequality in 1972.[4] They measured entangled photons, similar to those discussed in this chapter.

Freedman derived a Bell inequality that's even simpler than the original. We'll look only at the result here; as in

Einstein insisted on realism to preserve locality: the photons must have shared properties all along—from the moment they are created in a single location.

chapter 2, the proof is a bit too mathematical for this book. In chapter 4, though, we'll carefully develop additional Bell inequalities, without skipping any steps.

Freedman asks us to perform the following simple arithmetic:

• Record the number of photon pairs detected when the angle between the polarizers is 22.5° (figure 9a).

• From this, subtract the number of photon pairs detected when the angle between the polarizers is 67.5° (figure 9b).

• Multiply this result by 4.

• Is this greater than the number of photon pairs detected when the polarizers are removed? If so, then we've contradicted the common-sense assumption of local realism.

a. The angle between the polarizers is 22.5°.

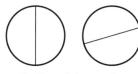

b. The angle between the polarizers is 67.5°.

Figure 9 Two of the three measurements required for a test of Freedman's inequality. The third measurement is made when the polarizers are removed.

To confirm that Freedman's rules are just simple arithmetic, let's look at a set of data. Freedman and Clauser didn't publish their raw data, so it will be convenient to look at data taken by my student, Charlotte Selton.[5] Detecting photon pairs over intervals of 50 seconds, she obtained

• 1821 photon pairs when the angle between polarizers was 22.5°,

• 377 photon pairs when the angle between polarizers was 67.5°, and

• 4474 photon pairs when the polarizers were removed.

Let's follow Freedman's rules: We start with 1,821 (corresponding with 22.5°), and then we subtract 377 (corresponding with 67.5°), which gives 1,444. Then we multiply this result by 4 to obtain 5,776. This is greater than 4,474, the number of pairs detected when polarizers were removed, so we've contradicted the constraint imposed by local realism!

Before we rigorously develop additional Bell inequalities, it will be useful to generalize a fact about the pairs of entangled photons described in this chapter. We've seen that:

• Both photons are vertical, or both are horizontal.

Equivalently:

- If one polarizer is horizontal and one polarizer is vertical, there's no possibility of a coincidence. (If one photon goes through a vertical polarizer, there's no possibility that the other photon goes through a horizontal polarizer.)

But what if we set the polarizers to some other angle? It turns out that our rule for the entangled photons can be generalized:

- If the angle between polarizers is 90°, there's no possibility of a coincidence. (If one photon goes through a polarizer, there's no possibility that the other photon goes through a perpendicular polarizer.)

This fact does not obviously follow from the method of creation of photon pairs: the original violet photon splits, producing a photon pair, in one of two crystals. One crystal produces vertically polarized photons, and the other produces horizontally polarized photons. Let's take it as an empirical fact that no coincidences occur when the polarizers are perpendicular to each other (regardless of whether the polarizers are horizontal, vertical, or diagonal). Another empirical fact is that just as many coincidences occur when both polarizers have the same angle (whatever the angle may be) as when both polarizers are horizontal, or both vertical.

Let's measure angles clockwise from the vertical, so vertical polarization is 0°, and horizontal is 90°. (Clockwise, of course, depends on which side of the polarizer we're looking at. We can choose either side—the side the photons are coming from, or the side they're going to—as long as we're consistent.) A negative angle represents counterclockwise rotation. We can now write two special cases of the rule above:

• There's no possibility of a coincidence when one polarizer is set to 30° and the other is set to 120°.

• There's no possibility of a coincidence when one polarizer is set to –30° and the other is set to 60°.

We'll return to these special cases in the next chapter. There, you will see for yourself the rigorous reasoning that forces physicists to reject local realism. You can decide for yourself whether the physicists' arguments make sense. Remember, local realism is what most of us would call common sense: objects have real properties that exist regardless of whether anyone is measuring them, and the measurement of one object can't affect a distant object. If this viewpoint is incorrect, even on the microscopic scale,

Local realism is what most of us would call common sense: objects have real properties that exist regardless of whether anyone is measuring them; the measurement of one object can't affect a distant object.

then something very weird is happening in our universe. (Or, as Philip Ball recently noted, *we're* weird for thinking that quantum physics is weird.[6] "Weird" means uncommon, but there's nothing more common than the particles that literally everything is made of—and these particles obey the laws of quantum physics.)

RIGOROUS CONTRADICTION OF EVERYDAY ASSUMPTIONS

Bell's original inequality is easy to articulate, as we saw in chapter 2. Freedman's version, the first Bell inequality tested in the lab, is even easier to put into words. But establishing those Bell inequalities is complicated. In this chapter, we'll look at simplified versions of Bell's theorem that can be proven rigorously, with very little math. The examples I chose for this chapter reveal some of my favorite quantum contradictions of common sense. I think it's instructive to study multiple examples because the reasoning is different in each case, but the conclusion is always the same. I find these examples more persuasive in the aggregate than in isolation.

Just Two Numbers, and then Two More: +1 and −1, +2 and −2.

Let's imagine an experiment performed with pairs of entangled photons.[1] Each photon travels toward an analyzer that determines whether the photon is polarized in a chosen direction. Let's imagine that the analyzer is a simple polarizer, followed by a detector to indicate whether each photon has passed through the polarizer.[2]

If a photon is transmitted through the polarizer, let's represent that result with the number +1. If it is not transmitted through the polarizer, we record −1. So to perform an experiment, we simply record +1 or −1 for each of the two photons.

We imagine two physicists, Alice and Bob, stationed at the two analyzers; each physicist takes responsibility for recording data at one of the analyzers. Each physicist dutifully records +1 or −1 for each photon arriving at the analyzer. Alice is interested in determining whether photons are polarized in the 0° direction (vertical). She's equally interested in determining whether the photons are polarized in the 45° direction. She sets her analyzer to either of these two angles. If she sets the angle to 0°, she uses the symbol A to represent the measured outcome (which is either +1 or −1). If she sets the angle to 45°, she uses the symbol A′ to represent the measured outcome (again, either +1 or −1).

There's no need to ruminate about why she chooses these two angles and not any others. The choice of these two angles is external to this discussion. We can think of the two angles simply as two settings of a switch.

Bob is also interested in two angles: 22.5° and 67.5°. When he sets his analyzer to 22.5°, he uses the symbol B to represent the measured outcome (+1 or –1). When he sets his analyzer to 67.5°, he uses the symbol B′ to represent the measured outcome (as always, either +1 or –1).

In effect, Alice is measuring A or A′, and Bob is measuring B or B′ (figure 10). The result of any measurement is either +1 or –1.

There are four possible combinations of measurements:

- Alice measures A and Bob measures B.

- Alice measures A and Bob measures B′.

- Alice measures A′ and Bob measures B.

- Alice measures A′ and Bob measures B′.

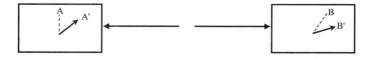

Figure 10 A simplified diagram of the measurement of a photon pair. Alice sets her analyzer to measure either A or A′, and Bob sets his analyzer to measure either B or B′. In this illustration, Alice is measuring A′, and Bob is measuring B′.

Next, we'll define a simple quantity, S. Let's say that S stands for Simple. S is calculated from A, A′, B, and B′:

$$S = AB + A′B - AB′ + A′B′.$$

S does not have any obvious physical meaning. It's just a quantity that we can calculate. We want to predict the possible values of S. For example, if A, A′, B, and B′ are all +1, we find that S is +2. If we change B to −1 while A, A′, and B′ are all still +1, S is −2. If we go through all possible combinations of values for A, A′, B, and B′, we find that S is always either +2 or −2.

Now, how can Alice and Bob measure S? For each photon pair, Alice measures either A or A′, not both, and Bob measures either B or B′, not both. So it's impossible to measure S for a single photon pair. Alice and Bob decide to take many measurements for all possible combinations of angles. So, when Alice measures A while Bob measures B, they obtain the product AB, and they find the average of this product over many measurements. Similarly they find the average of A′B over many trials in which Alice measures A′ while Bob measures B. They also find the average of AB′, and the average of A′B′. They obtain the average of every term in S, so they can calculate the average of S itself:

(average of S) = (average of AB) + (average of A′B) − (average of AB′) + (average of A′B′).

Now we make a key assumption. Even though Alice measures either A or A′, the photon is preprogrammed with both values: the photon has properties that predetermine the result of any possible measurement on it. This is Einstein's assumption of realism. Similarly, Bob's photon is preprogrammed with both values, B and B′, even though Bob measures only one of them for each photon. So, the four values A, A′, B, and B′ exist for each photon pair: they are hidden variables. So, the quantity S=AB+A′B−AB′+A′B′ exists for each photon pair.

Since S can only be −2 or +2, it's obvious that the average of S must be between −2 and +2. But experiment shows that the average of S is greater than 2! Where's the faulty reasoning that led to the false constraint on S?

Our mistake is the assumption that A, A′, B, and B′ all exist at the same time. Alice can only measure either A or A′; Bob can only measure either B or B′. The quantities that aren't measured don't have specific values that we can plug into S=AB+A′B−AB′+A′B′. So S doesn't exist for a single photon pair. Our belief that S should exist for a single photon pair is really our belief that photon properties exist before we measure them. Experiment contradicts this belief: the measured value of average S exceeds the limits imposed by realism.

All of the mathematical statements are really just abbreviations for statements about physical reality. Let's clarify this point. If we make just a single measurement of a photon, what does it mean to determine that A, for example, is +1? Does it mean that the number +1 was somehow imprinted on the photon, or riding along with the photon? Let's consider a more familiar kind of number: my weight, which is 150 pounds. Is the number 150 somehow imprinted on my body? But, isn't the pound an arbitrary unit? If I step on a metric scale, I see that I'm 68 kilograms. So which number am I carrying around with me? Is it 150, or 68, or something else? In fact, "150 pounds" and "68 kilograms" both represent a more fundamental fact: there are 40,000 trillions of trillions of protons and neutrons in my body. So "150 pounds" and "68 kilograms" are convenient, shorthand ways of specifying the total number of protons and neutrons in my body. (Electrons have mass too, but they're much lighter than protons and neutrons. All the electrons in my body weigh just about an ounce.) In a sense, the numbers 150 and 68 are indeed attached to me, because my weight is a very real quantity that can be measured any time. The numbers 150 and 68 represent factual statements about how much mass is in my body, regardless of whether I'm standing on a scale.

Similarly, the equation A = +1 is really just a code, or abbreviation, for a factual statement: this photon gets transmitted through Alice's polarizer if she sets it to 0°. Rather than saying, "This photon has a property that will cause it to pass through Alice's polarizer if she sets it to 0°," we just say "A = +1."

Experiment contradicts the assumption that every photon pair has values of A, A′, B, and B′. This exact same (and false) assumption can be written in a longer form: "The photon traveling toward Alice has properties that predetermine whether it will be transmitted or blocked by Alice's polarizer, regardless of whether she sets it to 0° or 45°. The photon traveling toward Bob has properties that predetermine whether it will be transmitted or blocked by Bob's polarizer, regardless of whether he sets it to 22.5° or 67.5°."

Did we assume locality, as well as realism, when we claimed that S must be +2 or −2? Yes, in a subtle way. We implicitly assumed that A, for example, is a property that doesn't depend on Bob's polarizer. In the definition of S— S=AB+A′B−AB′+A′B′—we assumed that A has the same value whether Bob measures B or B′.

Local realism imposes a constraint on a measurable quantity: the average value of S. Experiment violates this constraint, so at least one of our assumptions was wrong. If realism was the bad assumption, then the measurements do not reveal properties that the photons had all along; the photons were in some kind of undecided state prior to measurement. If locality was the bad assumption, then the angle of Bob's polarizer affects Alice's photons, and the angle of Alice's polarizer affects Bob's photons. Both of these possibilities are strange, and we'll see even stranger alternatives in chapter 6.

Perhaps you're curious about the single minus sign in $S=AB+A'B-AB'+A'B'$. When Alice measures A, her polarizer is set to 0°, and when Bob measures B', his polarizer is set to 67.5°. The angle between the two polarizers is 67.5°. In all other cases, the angle between the polarizers is 22.5°:

- AB: 0° on Alice's side and 22.5° on Bob's side.

- A'B: 45° on Alice's side and 22.5° on Bob's side.

- A'B': 45° on Alice's side and 67.5° on Bob's side.

The minus sign in S is associated with the one combination of angles that produces a 67.5° difference. The quantum prediction for the average of S is about 2.8, which violates the constraint imposed by local realism ($-2 \leq$ average $S \leq +2$). Experimental imperfections reduce the average of S below the ideal quantum prediction. The highest average S that I've measured is 2.66.

Do the Lights Match the Buttons?

Here's a variation on the same experiment, described in a book by Nicolas Gisin.[3] Alice and Bob install green and red lights on their analyzers. The green light flashes if the photon is polarized in the chosen direction; otherwise, the red light flashes. So, the green light corresponds with a +1 result in the previous example, and the red light corresponds with a −1 result.

Alice and Bob next install buttons to set the angles of the analyzers. Alice's buttons are labeled A and A′, and Bob's buttons are labeled B and B′. Each experimenter presses one button before the measurement of a photon pair. So, when Alice pushes the A button, she's measuring A, and when she pushes the A′, she's measuring A′. When Bob pushes the B button, he's measuring B, and when he pushes the B′ button, he's measuring B′. They decide to flip coins to decide which button to push so neither of them is influenced by the other person's choice. So, the four combinations of button presses are equally likely: A and B, A′ and B, A and B′, and finally A′ and B′.

Alice and Bob decide it's festive to paint their buttons, so they paint the A and B buttons green, and they paint the A′ and B′ buttons red (figure 11).

Now, Alice and Bob decide to award themselves points according to the following rules:

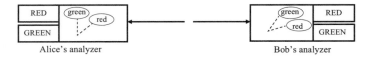

Figure 11 Alice and Bob install RED and GREEN lights on their analyzers, and red and green buttons.

• If Alice pushes her green button and Bob pushes his red button, they get a point if their lights flash **different** colors (one green, one red).

• For all other combinations of button presses, they get a point if their lights flash the **same** color (both green or both red).

After many measurements, over all combinations of button presses, **they find that 85 percent of the time, they get a point**.

We want to figure out how the photons manage to award points 85 percent of the time. We assume that the photons don't have any advance notice about which buttons will be pressed; perhaps the buttons are pressed while they are in flight. We also assume one photon can't send messages to the other about which button was pushed at its end. So, each photon is isolated from the other when it's compelled to cause either the green light or red light to flash.

So, the only information available to each photon is the direction of the analyzer that it encounters. The press of a red or green button sets the direction. What strategies are available to each photon? We can think of only four:

1. It can make the green light flash no matter which button was pressed.

2. It can make the red light flash no matter which button was pressed.

3. It can cause a flash of the light with the same color as the button that was pressed.

4. It can cause a flash of the light with a different color than the button that was pressed.

The photon could decide to randomly flash either the green light or the red light, but this is equivalent to randomly choosing between Strategy 1 and Strategy 2. So, the four strategies, including any kind of random selection among them, form a comprehensive set of options.

It's not necessary for both photons in a pair to be encoded with the same strategy; for example, one photon may follow Strategy 1, while the other follows Strategy 2. So, there are sixteen possible combinations of strategies for each photon pair: for each of the four choices available to one photon, the same four choices are available to the other photon. I'll go through just four of the sixteen combinations, the cases in which both photons follow the same strategy.

Suppose both photons follow Strategy 1, so that both lights flash green no matter which buttons are pressed. How often do Alice and Bob get a point? If both lights flash the same color, Alice and Bob get a point *unless* Alice presses her green button and Bob presses the his red button—and this happens 25 percent of the time because each of the four combinations of button presses is equally likely. So, Alice and Bob get a point 75 percent of the time if both photons employ Strategy 1. Yet, in fact, Alice and Bob get a point 85 percent of the time. So, Strategy 1, when employed by both photons, is inconsistent with experiment.

If both photons follow Strategy 2, both lights flash red no matter what, and again Alice and Bob get a point 75 percent of the time (in all cases except when Alice presses her green button and Bob presses his red button). So Strategy 2 isn't any more successful.

If both photons follow Strategy 3, the light that flashes is the same color as the button pressed. So let's examine all four combinations of equally likely button presses: When Alice and Bob both press their green button, both lights flash green, and they get a point. When both press their red button, both lights flash red, and they get a point.

When Alice presses green and Bob presses red, the lights flash different colors, but they get a point because this is the one combination of button presses in which a point is awarded for flashing two different colors. But when Alice presses red and Bob presses green, the lights flash different colors, and they don't get a point. So a point is awarded 75 percent of the time, and again the strategy is unsuccessful at reproducing experimental results.

Finally, if both photons follow Strategy 4, the light that flashes is different from the color of the button that was pressed. When Alice and Bob both press green, both lights flash red, and they get a point. When they both press red, both lights flash green, and they both get a point. When Alice and Bob press buttons of different colors, the lights flash different colors. This results in a point when Alice presses green and Bob presses red, but not when Alice presses red and Bob presses green. Once again, they get a point during only three of the four combinations of button presses.

We can also look at cases in which the two photons follow different strategies. But we'll never find a combination of strategies that awards a point more than 75 percent of the time. So no combination of strategies can produce the experimental result that a point is awarded 85 percent of the time.

This means we made at least one false assumption. We assumed locality: each photon is uninfluenced by the measurement made on the other photon. We also assumed realism: each photon has properties (a "strategy") that predetermine that outcome of any possible measurement; in other words, we assumed the measurement reveals a property that the photon already had.

Identical Twins: An Analogy

We'll establish another bona fide Bell inequality, without really using any math at all. All we need is basic logic. This example comes from Anton Zeilinger, who drew on earlier examples from Bernard d'Espagnat and Eugene Wigner.[4]

First, let's imagine a number of people in an auditorium. Some of the people, perhaps, have brown hair, and some, perhaps, have brown eyes. Let's compare the number of people with brown hair and brown eyes, to the number of people with brown hair and any color eyes:

(# with brown hair and brown eyes) ≤ (# with brown hair).

The left side indicates the number of people with brown hair and brown eyes. The right side indicates the number of people with brown hair and any eye color. The left side is more restrictive. The right side is larger than the left side by the number of people with brown hair and non-brown eyes. The number of people with brown hair and non-brown eyes may be 0, which is why we use ≤ instead of <. We're about to use this kind of logic to derive a Bell inequality.

Now we want to think about identical twins, subject to the following conditions:

- Both twins in a pair have the same height: tall or short.

- Both twins in a pair have the same hair color: brown or blond.

- Both twins in a pair have the same eye color: brown or blue.

We'll restrict our attention to twins with traits listed above. So, for example, we'll exclude twins with red, gray, or white hair, or no hair. We'll also assume that the twins don't dye their hair (or if they do, they dye it identically, either brown or blond).

According to these conditions, tall, brown-haired twins have either brown or blue eyes:

(# pairs tall, brown hair) = (# pairs tall, brown hair, brown eyes) + (# pairs tall, brown hair, blue eyes).

Let's establish the fact that

(# pairs tall, brown hair, brown eyes) ≤ (# pairs with brown hair, brown eyes).

This is true because the right side is less restrictive. The right side, but not the left, includes any pairs of brown-haired, brown-eyed twins who are short. Let's also use the fact that

(# pairs tall, brown hair, blue eyes) ≤ (# pairs tall, blue eyes).

This is true for a similar reason: the right side is less restrictive and includes any pairs of tall, blue-eyed twins who have blond hair. We can combine these two facts with the bold-faced equation above. Each term in the bottom line of the bold-faced equation will be replaced by a term that is at least as large:

(# pairs tall, brown hair) ≤ (# pairs with brown hair, brown eyes) + (# pairs tall, blue eyes).

The left side of this inequality indicates the number of pairs of twins who are tall and brown-haired, regardless of eye color. The tall, brown-haired twins with brown eyes are included in the first condition on the right side of the inequality, and the tall, brown-haired twins with blue eyes are included in the final condition. Thus, every pair of twins counted on the left side is counted once on the right side. The right side includes two additional sets of twins: any brown-haired, brown-eyed twins who are short, and any tall, blue-eyed twins who have blond hair. This is why the right side may be larger than the left.

We recognize that both twins are tall if one twin is tall, and both twins have brown hair if one twin has brown hair. When I write, "one twin is tall," I don't mean that *only* one twin is tall. Instead, I mean we might observe that one twin is tall, and we immediately infer that the other twin must also be tall. Now we can write

(# pairs tall, brown hair) = (# pairs in which one twin is tall and the other twin is brown-haired),

or, more succinctly,

(# pairs tall, brown hair) = #(one tall, one brown-haired).

Similarly,

(# pairs with brown hair, brown eyes) = #(one brown-haired, one brown-eyed),

and

(# pairs tall, blue eyes) = #(one tall, one blue-eyed).

Rewriting our bold-faced inequality this way, we obtain

#(one tall, one brown-haired) ≤ #(one brown-haired, one brown-eyed) + #(one tall, one blue-eyed).

This is a Bell inequality for pairs of twins. Its validity seems incontrovertible. But how does it apply to entangled photons? We saw in chapter 3 that we can create entangled photon pairs such that:

• If one polarizer is horizontal and one polarizer is vertical, there's no possibility of a coincidence. (If one photon goes through a vertical polarizer, there's no possibility that the other photon goes through a horizontal polarizer.)

Vertical polarization and horizontal polarization are *mutually exclusive*, like brown hair and blond hair (in our simple dichotomy). We can think of mutually exclusive hair colors as analogues of mutually exclusive polarization directions. Let's associate vertical polarization with brown hair, and horizontal polarization with blond hair. What about eye color, and height? We'll associate them with other mutually exclusive polarizer angles that we saw in the previous chapter:

• There's no possibility of a coincidence when one polarizer is set to 30° and the other is set to 120°.

• There's no possibility of a coincidence when one polarizer is set to −30° and the other is set to 60°.

We'll associate 30° with tall, and 120° with short. Similarly, we'll associate −30° with blue eyes, and 60° with brown eyes. Here's the summary of polarizer angles and their associated traits:

0° (vertical): brown hair

90° (horizontal): blond hair

30°: tall

120°: short

–30°: blue eyes

60°: brown eyes

Now we simply translate our Bell inequality for twins,

#(one tall, one brown-haired) ≤ #(one brown-haired, one brown-eyed) + #(one tall, one blue-eyed),

into a Bell inequality for entangled photons:

#(one 30°, one 0°) ≤ #(one 0°, one 60°) + #(one 30°, one –30°).

Rewriting this more compactly,

$$N(30°,0°) \leq N(0°,60°) + N(30°, -30°),$$

where $N(30°,0°)$ is the number of coincidences measured in some time interval when one polarizer is set to 30°, and the other is set to 0° (vertical). $N(0°,60°)$ and $N(30°,-30°)$

are defined similarly. This inequality is violated by measured data!

The Bell inequality, when applied to twins, is absolutely irrefutable. If you reread the steps leading to the result, I hope you'll agree that there are no dubious assumptions at all. If you actually do a survey of twins (who are either tall or short, either brown-haired or blond-haired, and either brown-eyed or blue-eyed), the inequality will certainly be satisfied. So why does the inequality fail for entangled photons? We have to think carefully about the (very reasonable) assumptions that we made about twins.

We assumed realism: eye color, for example, is completely independent of whether anyone is observing it. (The role of observation is not explicit in our derivation of the inequality. But before counting up pairs of twins with certain combinations of traits, someone has to observe those traits.) Prior to observation, the twin's eye color is not some indeterminate mixture of brown and blue, which instantly coalesces to one color or the other at the moment of observation. Indeed, the DNA, shared by both twins, is the hidden variable that causes identical twins to have identical traits regardless of whether anyone is observing them. (DNA is the hidden variable as long as the twins don't dye their hair or wear colored contact lenses.)

We also assumed locality: one twin's height does not depend on whether the other twin's height, hair, or eyes are being observed! This assumption is so obviously true that we're not even aware that we're making it. Yet this assumption, combined with the assumption of realism, leads to the Bell inequality that definitely does *not* apply to entangled photons.

When we rewrote the inequality to apply to entangled photons, we implicitly assumed that every photon all along has properties that predetermine whether it will pass through a polarizer at any chosen angle (just as the twins all along have definite eye color, hair color, and height). Indeed, the photons would have to *satisfy* the inequality if they had fixed polarization properties all along (realism), and if each photon was uninfluenced by the other photon's polarizer (locality). But in fact, the entangled photons *violate* the inequality.

So, we've done it again! Assuming local realism, we derived a constraint that is violated by measurements of entangled photons. We used practically no math at all, just pure logic. Why, then, do physicists usually use Bell inequalities based on calculus? Are physicists being deliberately difficult to confuse outsiders and make themselves seem smart?

I think there's a legitimate reason why other Bell inequalities are valuable. In the derivations of other Bell inequalities, the assumptions of realism and locality appear very explicitly in specific steps. The derivation above, though as simple as we can hope for, somewhat obscures

Indeed, the photons would have to *satisfy* the inequality if they had fixed polarization properties all along (realism), and if each photon was uninfluenced by the other photon's polarizer (locality).

the exact role of the assumptions (which we're not even aware we're making when thinking about twins).

Let's consider the source of entangled photons described in chapter 3: half the photon pairs are horizontally polarized, and half are vertically polarized. Let's think a little about the quantum prediction for the three terms in the inequality,

$$N(30°,0°) \leq N(0°,60°) + N(30°, -30°).$$

For our entangled photons, the quantum prediction (confirmed by measurement) is that the term on the left side is 50 percent larger than the sum on the right. We can actually prove this result using just a few ideas:

• As stated in chapter 3, a photon has a 75 percent chance of passing through a polarizer tilted 30° from the photon's polarization. And, a photon has a 25 percent chance of passing through a polarizer tilted 60° from the photon's polarization.

• Assume one photon is measured slightly (or much) earlier than the other. The first photon has a 50 percent chance of passing through the polarizer (regardless of its orientation). This is a fact that we haven't previously stated, but it's true.

• Let's suppose, just for convenience, that the measurement of one photon immediately creates a definite polarization for both photons. This is spooky action at a distance: when one photon passes through a polarizer, the distant photon immediately acquires the same polarization (for all practical purposes). Some physicists strenuously reject spooky action at a distance, which is why I include the disclaimer "for all practical purposes." (We will discuss interpretations of quantum physics in chapter 6.)

To predict N(30°,0°), N(0°,60°), and N(30°, −30°), we need to think about the efficiency of the polarizers and detectors. It's possible to buy very good polarizers that behave almost ideally: they let through all the photons they're supposed to, and they block all the photons they're supposed to. Single-photon detectors, though, usually have efficiencies much less than 100 percent: they fail to respond to some of the photons that reach them. But detector efficiency affects every coincidence count by the same factor. If all the terms in our Bell inequality are reduced by the same factor, there's no change in which side is larger. So we'll ignore detector efficiency and briefly revisit it in chapter 6.

Let's calculate N(30°,0°), the number of coincidences detected when one polarizer is set to 30°, and the other is set to 0°. Suppose the first photon reaches the 30° polarizer. The photon has a 50 percent chance of passing through because the first photon in a pair to reach a polarizer (at any angle) has a 50 percent chance of passing through. If it passes through, the other photon acquires the 30° polarization and thus has a 75 percent chance of passing through the 0° polarizer. Consequently, the probability that a photon pair passes through both polarizers is 50% × 75% = 37.5% = $\frac{3}{8}$. In other words, when one polarizer is set to 30° and the other to 0°, $\frac{3}{8}$ of the photon pairs pass through both polarizers. If the total number of photon pairs encountering the polarizers is N_{total}, then $N(30°,0°) = \frac{3}{8}N_{total}$.

Let's do the same for $N(0°,60°)$. If the first photon encounters the 0° polarizer, it has a 50 percent chance of passing through. If it passes through, the other photon acquires the 0° polarization and thus has a 25 percent chance of passing through the 60° polarizer. (The angular difference between the polarizers is now 60°; it was 30° in the previous case.) Thus, the probability that a photon pair passes through both polarizers is 50% × 25% = 12.5% = $\frac{1}{8}$. In other words, when one polarizer is set to 0° and the other to 60°, $\frac{1}{8}$ of the photon pairs pass through both polarizers, and $N(0°,60°) = \frac{1}{8}N_{total}$. So $N(0°,60°)$ is $\frac{1}{3}$ of $N(30°,0°)$ because there are fewer coincidences when there's a greater difference between polarizer angles.[5]

N(30°,–30°) is the same as N(0°,60°) because the angular difference between the two polarizers is 60° in both cases. The probability that a photon pair passes through both polarizers depends only on the angular difference between the two polarizers. Thus, the coincidence count also depends only on the difference between the two polarizers, and $N(30°,–30°) = N(0°,60°) = \frac{1}{8}N_{\text{total}}$.

We now have explicit expressions for all three terms in our Bell inequality:

$N(30°,0°) \leq N(0°,60°) + N(30°, –30°),$

which becomes

$\frac{3}{8}N_{\text{total}} \leq \frac{1}{8}N_{\text{total}} + \frac{1}{8}N_{\text{total}}.$

This simplifies to

$\frac{3}{8} \leq \frac{2}{8},$

which is clearly untrue. Experiment shows that the coincidence count on the left side is indeed 50 percent larger than the sum on the other side, confirming quantum mechanics and overruling local realism.

Everyday observations, in our familiar macroscopic world, *do* conform to local realism. If we actually performed the observations of pairs of twins, the Bell inequality for twins would be satisfied, not violated. We can imagine how the twins would have to behave to conform to the quantum predictions. We have to imagine that prior to observation, the twins' height, hair color, and eye color are undetermined and somehow *undecided*, not merely unknown. Before either twin is observed, there's a 50 percent probability that a twin will be tall once height is observed; likewise, there is a 50 percent probability that a twin will be short. Similarly, brown hair and blond hair are equally likely, and brown eyes and blue eyes are equally likely.

Suppose the twins' names are Jordan and Pat. If Jordan is observed to be tall, then Pat, for all practical purposes, immediately becomes tall. Even more strangely, once Pat's hair or eye color is observed, Pat's height reverts to being unknown, and there's a chance that a subsequent observation will reveal Pat's height to be *short* (no longer tall like Jordan)!

This last effect is indeed true for entangled photons: as soon as the polarization of one photon is observed, the polarization of the other is known to be the same.[6] But no further measurement of either photon affects the other. The entanglement is *severed* the instant it takes effect, exactly when a measurement of *one* photon creates a definite polarization for *both*. This provides a possible explanation for why we don't observe entanglement on a macroscopic scale: the very many particles making up macroscopic objects are constantly interacting and in some sense "measuring" one another, thereby disentangling as fast they entangle.[7]

The everyday assumption of local realism is contradicted by observations of entangled photons. Moreover, if macroscopic objects started behaving like entangled photons, we would be assailed by a host of bizarre effects. We'll look at other constraints imposed by local realism—constraints that are violated by measured results.

Three Observations by Maudlin

This next example is from *Quantum Non-Locality and Relativity* by the philosopher Tim Maudlin.[8] We'll be thinking about photons all along, and making the assumption of realism more explicitly than in the previous example. The assumption of locality will be implicit, at first.

We'll continue to work with the pairs of entangled photons described previously (each pair is 50 percent likely to pass through horizontal polarizers, and 50 percent likely to pass through vertical polarizers). We now specify that each photon will encounter a polarizer set to one of three angles: vertical, 30° from the vertical (in the clockwise direction), and 60° from the vertical (in the clockwise direction). As we might predict from our previous discussion, the following three observations can be made:

Observation 1. If the two polarizers have the same orientation, the two photons always do the same thing: either both photons pass through the polarizers, or both are blocked by the polarizers.

Observation 2. If the angle between the two polarizers is 30°, the two photons do the same thing 75 percent of the time; 25 percent of the time, one passes through its polarizer, while the other is blocked.

Observation 3. If one of the polarizers is vertical and the other is at 60°, the two photons do the same thing 25 percent of the time.

This is based on real observations; this isn't made up.

Let's think about Observation 1. If the two polarizers are in the same direction, the two photons always do the same thing. How do they accomplish this? It must be, according to common sense, that the two photons share a common property. It seems intuitively obvious that the photons have this common property all along—from the moment they're created. Our common sense and intuition are based on realism: the photons have "hidden properties" prior to observation; observation merely lets us view the properties that the photons had all along. Let's see where this assumption leads.

The hidden properties of a photon might be this:

(Would pass through a vertical polarizer. Would be blocked by a 30° polarizer. Would pass through a 60° polarizer.)

It took a lot of words to write that. Let me represent *exactly the same thing* in abbreviated form:

(Vertical→Pass. 30°→Block. 60°→Pass.)

If one photon in a pair has these hidden properties, the other one must have the same properties. To confirm this, suppose one photon in a pair has the hidden properties listed above, and the other one has these properties:

(Vertical→Block. 30°→Block. 60°→Pass.)

This says that if both polarizers are vertical, the two photons do different things: one passes through the polarizer, and one is blocked. This contradicts Observation 1, which requires the two photons to do the same thing if the polarizers are set to the same angle (in this case, vertical). So the two photons in a pair must share the same set of hidden properties.

These hidden properties might be thought of as instruction sets, telling the photons what to do when they reach polarizers. Or, the hidden properties may be thought of as features that are detected by polarizers at the appropriate angles. I like to think of the hidden properties as tickets, and the polarizers are bouncers, who admit only the bearers of appropriately marked tickets. In all cases, the hidden properties are qualities inherent in the photons all along, prior to measurement. This is realism.

We actually already made the assumption of locality as well: we assumed that what a photon does at a polarizer depends only on the angle of *that* polarizer. We assumed that a photon's ability to pass through a polarizer does *not* depend on the angle of the polarizer that the *other* photon encounters. This assumption seems so natural that we have to go out of our way to recognize that we're making it. Now, we'll see where the assumptions of realism and locality lead us.

Let's consider again the first example of hidden properties:

(Vertical→Pass. 30°→Block. 60°→Pass.)

If all photons had exactly these properties, the two photons in a pair would always do the same thing (pass) whenever one polarizer was vertical, and the other was 60°. But this should happen only 25 percent of the time, according to Observation 3. So a certain fraction of photons might have the hidden properties listed above, but other photons must have different hidden properties.

Now, let's list all possible sets of hidden properties, in four groups. Two sets of hidden properties are in each group.

Group 1: The two photons do the same thing, regardless of the orientation of the polarizers.

(Vertical→Pass. 30°→Pass. 60°→Pass.): each photon always passes.

(Vertical→Block. 30°→Block. 60°→Block.): each photon is always blocked.

Group 2: The two photons do the same thing, unless exactly one of the polarizers is vertical.

(Vertical→Block. 30°→Pass. 60°→Pass.): each photon passes unless it reaches a vertical polarizer.

(Vertical→Pass. 30°→Block. 60°→Block.): each photon is blocked unless it reaches a vertical polarizer.

Group 3: The two photons do the same thing, unless exactly one of the polarizers is 30°.

(Vertical→Pass. 30°→Block. 60°→Pass.): each photon passes unless it reaches a polarizer at 30°.

(Vertical→Block. 30°→Pass. 60°→Block.): each photon is blocked unless it reaches a polarizer at 30°.

Group 4: The two photons do the same thing, unless exactly one of the polarizers is 60°.

(Vertical→Pass. 30°→Pass. 60°→Block.): each photon passes unless it reaches a polarizer at 60°.

(Vertical→Block. 30°→Block. 60°→Pass.): each photon is blocked unless it reaches a polarizer at 60°.

We've listed all eight possible sets of hidden properties. This is a complete list: for each of the three polarizer angles, two outcomes are possible (pass or block), and we listed all possible combinations of outcomes.

Our final task is to determine the fraction of photon pairs that have hidden properties from each of the four groups. Let F_1 be the fraction (between 0 and 1) of photon pairs whose hidden properties are from Group 1. Then F_2, F_3, and F_4 are similarly defined.

Now, let's consider Observation 3, based on one vertical polarizer and one 60° polarizer. Let's imagine leaving the polarizers at these angles for a long time. Many photon pairs encounter the polarizers, and the two photons do the same thing 25 percent of the time, as also stated in Observation 3. So exactly 25 percent of the photon pairs must be from groups that do the same thing when the polarizer angles differ by 60°. When we look over the four groups, we see that photon pairs in Group 1 and also Group 3 do the same thing when the polarizer angles differ by 60°: photon pairs from Group 1 always do the same thing, and photon pairs from Group 3 do the same thing as long as one angle isn't 30°. (In contrast, the photons in Groups 2 and 4 do *different* things when the polarizer angles differ by 60°: one photon passes, and one is blocked.) So 25 percent of the photons pairs must be from either Group 1 or Group 3. We don't yet know how the 25 percent is divided; it's not necessarily 12.5 percent from Group 1 and 12.5 percent from Group 3. We just know that the total fraction of photon pairs from Groups 1 and 3 combined is 25 percent. We can write this fact mathematically as:

$F_1 + F_3 = 0.25.$

Next, we'll apply the same logic to Observation 2. Let's consider Observation 2, in the specific case of one vertical polarizer and one 30° polarizer (a 30° difference). We imagine leaving the polarizers set to these angles for a long time, so that many photon pairs encounter these angles. According to Observation 2, the two photons in a pair do the same thing 75 percent of the time. In other words, 75 percent of the photon pairs are from groups that do the same thing when one polarizer is vertical and the other is at 30°. We find that photon pairs from Group 1 always do the same thing, and photon pairs from Group 4 do the same thing as long as one polarizer isn't 60°. (Photon pairs from the two other groups do different things when one polarizer is vertical and the other is at 30°.) This means 75 percent of the photons pairs are from Groups 1 and 4:

$F_1 + F_4 = 0.75.$

We repeat for the other case of Observation 2, with one 30° polarizer and one 60° polarizer (still a 30° difference). Again we imagine leaving the polarizers set this way for a long time. The two photons do the same thing 75 percent of the time, so 75 percent of the photons must be from groups that do the same thing when one angle is 30° and the other is 60°. We find this behavior in Groups 1 and 2. Therefore, 75 percent of photon pairs must be from Groups 1 and 2:

$F_1 + F_2 = 0.75.$

Finally, we use the fact that the sum of all fractions is 1 (100 percent of the photon pairs come from one of the four groups):

$F_1 + F_2 + F_3 + F_4 = 1.$

If you subtract the first three equations from the last one, you find

$F_1 + F_2 + F_3 + F_4 - (F_1 + F_3) - (F_1 + F_4) - (F_1 + F_2) =$
$1 - 0.25 - 0.75 - 0.75.$

This simplifies to

$-2F_1 = -0.75$

or

$F_1 = 0.375$.

Plugging this into the first equation, $F_1 + F_3 = 0.25$, yields $F_3 = -0.125$. But we can't have a negative fraction of photons! Therefore at least one of our assumptions is disproven. More specifically, either realism is false, or locality is false, or both are false.

If realism is false, the photons do *not* have their observed properties prior to observation; prior to observation, the photons are in a mysterious indeterminate state. If realism is false, the measurement of the photons definitely changes something. But what changes? An objective physical property of the photons, or just our knowledge of the photons? Does measurement create objectively real states, or is direct observation the only reality accessible to science? These questions are answered differently in different interpretations of quantum mechanics. There is no consensus among physicists.

There is a way to save realism: we can discard locality. In this case, each photon's behavior depends on the angle of *both* polarizers: the polarizer it encounters, and the polarizer encountered by the distant photon. But polarizers can be made out of sheets of plastic. Why should a photon be affected by a distant piece of plastic? The only thing linking the photon to that piece of plastic is the fact that the *other* photon encounters it. I find it especially spooky to imagine one photon being affected by the other photon's polarizer.

Or, we can discard both realism and locality. This permits an interpretation of quantum mechanics that I find simple and expedient: measurement creates objectively real states. Because entangled photons share a single state, the measurement of one photon immediately creates an objectively real state for both photons. This interpretation is nonlocal because the measurement of one photon physically alters the distant photon (for all practical purposes). Although this interpretation allows for objective reality, the objectively real state is created by the measurement and does not exist prior to the measurement. Thus, this interpretation is not consistent with realism, which requires measurable properties to exist prior to measurement.

A further sampling of interpretations of quantum mechanics appears at the end of the book.

Hardy Makes It Easy

We'll take a look now at an example that doesn't require us to think of *proportions* of photon pairs doing one thing or another. We only need to know that some outcomes never occur, and some outcomes occasionally occur. Lucien Hardy came up with the basic idea for this example, but I'm following a version by N. David Mermin.[9]

Consider entangled photon pairs from a certain source. These photons are entangled differently from the cases we've seen previously; they're not 50 percent horizontally polarized and 50 percent vertically polarized. We aren't concerned with the specifics of the new entangled state. We need to know only that this new state can be created experimentally.

The two photons in a pair travel in different directions, each to its own analyzer. There's a switch on the analyzer that allows the experimenter to select one of two settings: Setting 1 or Setting 2. All we need to know is that the analyzers are wired to lights: a green light and a red light. Each time a photon arrives at an analyzer, either the green light flashes, or the red light flashes (figure 12).

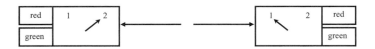

Figure 12   The two photons in a pair go their separate ways, each to an analyzer. The analyzer has two settings, Setting 1 and Setting 2, and two lights, red and green. Each time a photon arrives at an analyzer, one of the lights flashes.

After recording a large number of results for all possible combinations of switch settings, we make the following observations:

Observation A. When both switches are set to Setting 1, at least one light flashes green.

Observation B. When one switch is set to Setting 1 and the other is set to Setting 2, at least one light flashes red.

Observation C. When both switches are set to Setting 2, occasionally both lights flash green.

How do the photon pairs manage to comply with these rules? One possibility is that the two photons are in communication with each other: spooky action at a distance, which is a familiar concept by now. But we'll stubbornly cling to our cherished common-sense notion of local realism. We'll attempt to explain the photons' behavior under the assumption that each photon all along has a hidden property that is detected by the analyzer, independent of the other photon or the other photon's analyzer.

Since there are only two switch settings, there are only four possible sets of hidden properties for each photon:

(Setting 1→Green. Setting 2→Green.)

(Setting 1→Green. Setting 2→Red.)

(Setting 1→Red. Setting 2→Green.)

(Setting 1→Red. Setting 2→Red.)

Equivalently, we can say that each photon has one property from each of the two columns below:

Setting 1 Column	Setting 2 Column
Setting 1→Green.	Setting 2→Green.
Setting 1→Red.	Setting 2→Red.

Unlike previous cases, it is *not* necessary that the two photons in a pair share the same hidden properties. In fact, it's clear from Observation A that if one photon causes a red flash when its analyzer is set to Setting 1, then the other photon must have a different property. So we need to figure out which properties one photon may or may not have, based on the other photon's properties.

Let's look at Observation B first. Since there's at least one red flash when the switch settings differ from each other, if one light flashes green on Setting 1, the other must flash red on Setting 2. In other words, if one photon has the (Setting 1→Green) property, the other must have the (Setting 2→Red) property. Similarly, if one photon has the (Setting 2→Green) property, the other must have the (Setting 1→Red) property. Thus whenever both photons have the (Setting 2→Green) property (which occasionally happens, according to Observation C), both photons must also have the (Setting 1→Red) property. But this contradicts Observation A because then both lights would flash red when both Settings are 1!

That argument goes by fast, so let's decompress it. We'll now approach the situation as if it were a logic puzzle in which you have to figure out where to seat people at a table, given that Mr. A insists on sitting next to Ms. B but at least two seats away from Dr. C, and so on. In our case, each photon has one of the four sets of hidden properties, and the other photon also has one of the four sets. Thus, there are sixteen different combinations. I'll represent this in the table below, where the rows represent the hidden properties of one photon (which I'll call the row photon), and the columns represent the hidden properties of the other photon (which I'll call the column photon). G and R stand for green and red.

	1→G. 2→G.	1→G. 2→R.	1→R. 2→G.	1→R. 2→R.
1→G. 2→G.				
1→G. 2→R.				
1→R. 2→G.				
1→R. 2→R.				

We can eliminate some of these sixteen possible combinations by studying the three observations. Observation A is that if both switches are set to 1, at least one light flashes green. So, we can eliminate the cases in which both switches are set to 1 and both lights flash red. I'll mark these forbidden cases in the table with an A, indicating that Observation A forced us to eliminate them:

	1→G. 2→G.	1→G. 2→R.	1→R. 2→G.	1→R. 2→R.
1→G. 2→G.				
1→G. 2→R.				
1→R. 2→G.			A	A
1→R. 2→R.			A	A

Just to be clear why we did this, let's look at the A in the bottom right corner (last column, last row), where both photons have the hidden property (1→R. 2→R.) If both switches are set to 1, both lights flash red. This violates Observation A. The other three cases where A appears in the table occur where both photons have the property 1→R.

Now let's look at Observation B: If the switches have different settings, at least one light flashes red. Now we can eliminate all the cases where both lights flash green when the switch settings are different. I'll mark these cases with a B:

	1→G. 2→G.	1→G. 2→R.	1→R. 2→G.	1→R. 2→R.
1→G. 2→G.	B	B	B	
1→G. 2→R.	B		B	
1→R. 2→G.	B	B	A	A
1→R. 2→R.			A	A

Let's clarify what we just did. Start with the B in the upper left. In this case, both photons have the hidden property (1→G. 2→G.). No matter what, both lights flash green. They flash green when the switch settings are the same; more significantly, they also flash green when the switch settings are different, in violation of Observation B. Next let's move one space to the right in the table, where the row photon still has the property (1→G. 2→G.), and the column photon has the property (1→G. 2→R.). The row photon always causes its analyzer to flash green. If the column photon's analyzer is set to 2, then there's a red flash on the column photon's side. This isn't a violation of Observation B. But if the row photon's analyzer is set to 2, and the column photon's analyzer is set to 1, both lights flash green. This does violate Observation B, so we have to forbid this combination of hidden properties.

Similar reasoning leads to the other appearances of B in the table: any time there is a possibility of two green flashes occurring when the switches are different, we have to forbid it.

Using Observations A and B, we've eliminated eleven combinations of hidden properties, but five still remain. Can any of these five combinations satisfy Observation C? Observation C requires some photon pairs to cause two green flashes when both switches are set to 2. In other words, in some pairs, both photons must have the 2→G property. There are four combinations in the table such that both photons have the 2→G property. I'll mark them with an asterisk (*):

	1→G. 2→G.	1→G. 2→R.	1→R. 2→G.	1→R. 2→R.
1→G. 2→G.	B*	B	B*	
1→G. 2→R.	B		B	
1→R. 2→G.	B*	B	A*	A
1→R. 2→R.			A	A

All four combinations that could satisfy Observation C are strictly forbidden, either by Observation A or Observation B. Thus, local realism once again fails to explain our observations. And once again, we can ask, what alternative explanation can we give? We can attribute the results to spooky action at a distance: the two photons are in some kind of communication with each other, such that the measurement of one photon influences the measurement of the other. Or, we can assert that observation is the only scientific reality; we recognize that quantum predictions are invariably accurate, but we refuse to speculate about underlying causality. In other words, we acknowledge that local realism doesn't work, but we don't suggest any alternatives.

I survey additional interpretations of quantum physics in chapter 6.

Three Entangled Photons

Now we'll look at a case of three entangled photons: a trio discovered, appropriately, by a trio of physicists: Daniel Greenberger, Michael Horne, and Anton Zeilinger.[10]

Each of the three photons enters an analyzer similar to the analyzers of the previous example: each analyzer has a switch that can be set to Setting 1 or Setting 2, and every time a photon arrives, two outcomes are possible. Instead of flashing lights, we now have a digital display that shows either +1 or −1 (figure 13).

After the three photons enter the three analyzers, we record whether +1 or −1 is shown on each of the three displays. We repeat the experiment many times for

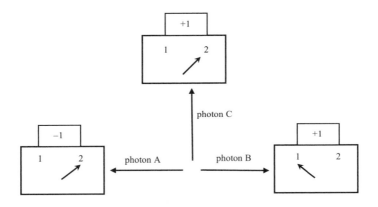

Figure 13 Three particles arise from a common source and separate, traveling to three analyzers. Each analyzer has a switch with two settings and a digital display that shows +1 or −1. This figure illustrates a possible combination of switch settings and digital results.

various combinations of switch settings on the three analyzers. For simplicity, we'll restrict our attention to cases in which an odd number of switches is set to Setting 1. In other words, either all switches are set to Setting 1, or one switch is set to Setting 1, and the other two are set to Setting 2.

We observe the following two facts:

Fact 1. When one switch is set to Setting 1 and the other two are set to Setting 2, −1 is shown either on one display or all three displays. (So +1 is shown on either two displays or no displays.) Since an odd number of displays shows −1, the product of the three displayed numbers must be –1: if –1 is on one display, the product of the three numbers is $–1 \times 1 \times 1 = –1$; if –1 is on all three displays, the product is $–1 \times (–1) \times (–1) = –1$.

Fact 2. When all three switches are set to Setting 1, –1 is shown either on two displays or no displays. Since an even number of displays shows –1, the product of the three displayed numbers must be +1: if –1 is on two displays, the product of the three numbers is $–1 \times (–1) \times 1 = 1$; if –1 is on no displays, the product is $1 \times 1 \times 1 = 1$.

These observations are consistent with the quantum prediction, which we don't need to get into. Our purpose is to show that the observations are incompatible with the assumption of local realism.

As we've seen in previous examples, local realism implies that each photon carries within it hidden properties. The outcome of a measurement is determined by the photon's hidden properties and the switch setting at the analyzer that it enters; the photon isn't influenced by the other photons or the switch settings at their analyzers. Thus, each photon must have one of these four hidden properties:

Property I:	(Setting 1→+1. Setting 2→+1.)
Property II:	(Setting 1→+1. Setting 2→–1.)
Property III:	(Setting 1→–1. Setting 2→+1.)
Property IV:	(Setting 1→–1. Setting 2→–1.)

This is a complete list of the hidden properties a photon might have; there are two possible outcomes for each of the two possible switch settings.

It will be convenient to use symbols to represent the numbers displayed on the analyzers. Let the letters A, B, and C represent the three photons. Let's use a subscript to represent the setting of the switch. So, if Photon A's analyzer is set to Setting 1, A_1 represents the displayed number. If Photon A's analyzer is set to Setting 2, A_2 represents the displayed number. We can think of A_1 and A_2 as the hidden properties themselves, which now really are hidden variables. Thus, the four possible hidden properties of Photon A are the four possible combinations of values of A_1 and A_2:

Property I:	$(A_1 = +1. \ A_2 = +1.)$
Property II:	$(A_1 = +1. \ A_2 = -1.)$
Property III:	$(A_1 = -1. \ A_2 = +1.)$
Property IV:	$(A_1 = -1. \ A_2 = -1.)$

B_1 and B_2 are defined similarly for Photon B, and C_1 and C_2 are defined similarly for Photon C. We have two hidden variables for each photon, and there are three photons, so we have a total of six hidden variables. Each hidden variable has two possible values, so there's a total of

$2 \times 2 \times 2 \times 2 \times 2 \times 2 = 64$ combinations of values. Do we have to explicitly consider all sixty-four combinations?

Luckily, we can quickly show that none of the sixty-four combinations is consistent with the two facts. According to Fact 1, whenever one switch is set to Setting 1 and the other two are set to Setting 2, the product of the three displayed numbers is -1. We can look at the three possible cases, remembering that *the subscripts indicate the switch settings*:

Fact 1 (three cases):

- Photon A's analyzer is set to Setting 1, and the other two are set to Setting 2. This gives us $A_1B_2C_2 = -1$.

- Photon B's analyzer is set to Setting 1, and the other two are set to Setting 2. This gives us $A_2B_1C_2 = -1$.

- Photon C's analyzer is set to Setting 1, and the other two are set to Setting 2. This gives us $A_2B_2C_1 = -1$.

Let's see if the three equations above can tell us what $A_1B_1C_1$ is. We'll use the fact that 1^2 equals 1, and $(-1)^2$ also equals 1. Since each hidden variable (A_1, A_2, B_1, B_2, C_1, and C_2) is either 1 or -1, the square of any hidden variable is always 1. Thus we can write, for example, $A_2 \times A_2 = A_2{}^2 = 1$, $B_2 \times B_2 = B_2{}^2 = 1$, and $C_2 \times C_2 = C_2{}^2 = 1$.

We'll just multiply $A_1B_1C_1$ by 1 three times:

$A_1B_1C_1 = A_1B_1C_1 \times 1 \times 1 \times 1.$

We'll replace one 1 using $1 = A_2 \times A_2$, we'll replace another 1 using $1 = B_2 \times B_2$, and we'll replace the third 1 using $1 = C_2 \times C_2$:

$A_1B_1C_1 = A_1B_1C_1 \times A_2 \times A_2 \times B_2 \times B_2 \times C_2 \times C_2.$

Next, we'll just reorder the variables on the right:

$A_1B_1C_1 = A_1B_2C_2 \times A_2B_1C_2 \times A_2B_2C_1.$

The three terms on the right side are precisely the three expressions that came out of Fact 1. As we saw above, each of those three terms on the right is -1:

$A_1B_1C_1 = (-1) \times (-1) \times (-1) = -1.$

We thus conclude that $A_1B_1C_1 = -1$: when the three switches are all set to Setting 1, the product of the three displayed numbers is -1. But this contradicts Fact 2, which states that the product of the displayed numbers must be $+1$ when the three switches are set to Setting 1!

Facts 1 and 2 arise from quantum theory and are confirmed by measurement. But local realism, once again, fails to accommodate experimental facts. In fact, local realism predicts the exact opposite of Fact 2 (when Fact 1 is a given). Yet again, we are forced to discard the assumption that each photon, independently of the other

Each photon behaves as if it's monitoring the switch settings at all three analyzers, and colluding with the other two photons to satisfy the observed facts.

photons and their analyzers, is predestined to behave in a particular way when encountering a measuring device. Each photon behaves as if it's monitoring the switch settings at all three analyzers, and colluding with the other two photons to satisfy the observed facts.

RECONCILING WITH RELATIVITY

The two pillars of modern physics are relativity and quantum physics. These two fields of discovery unsettle not only our common sense but also seem even to unsettle *each other*. Let's examine Einstein's relativity to identify, and resolve, an apparent conflict with quantum entanglement.

The Shocking Truths of Einstein's Relativity

Physicists before Einstein recognized that light is an electromagnetic wave. In fact, the speed of light is hidden like a buried treasure within the laws of electricity and magnetism. These tried and true laws predict a very specific speed for light in empty space: 670 million miles per hour.

The two pillars of modern physics are relativity and quantum physics. These two fields of discovery unsettle not only our common sense but also seem even to unsettle *each other*.

Einstein thought very carefully about this fact. He recognized that the laws of physics must be the same for everyone, no matter where they are or how fast they're going. Specifically, the electromagnetic laws, which predict the speed of light, must be the same for everyone, no matter where they are or how fast they're going. This means that everyone, no matter how fast they're going, must measure the same speed for light in empty space. This single insight, apparently innocent, revolutionizes our understanding of space and time.

Imagine a baseball pitcher who can throw a ball at 80 miles per hour. Now the pitcher throws a ball at 80 miles per hour from a cart moving 40 miles per hour (figure 14). The ball is moving 80 miles per hour relative to the pitcher and the cart, so the ball is moving 120 miles per hour relative to the ground. This is common sense.

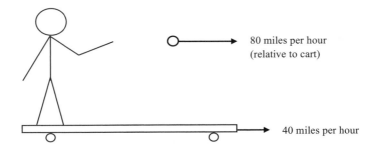

80 miles per hour
(relative to cart)

40 miles per hour

Figure 14 If a baseball pitcher rides a cart going 40 miles per hour and throws a baseball at 80 miles per hour relative to the cart, the baseball's speed is 120 miles per hour relative to the ground.

Now replace the cart with a spaceship going 335 million miles per hour, relative to Earth. Replace the baseball with light from the spaceship's headlight (figure 15). The light is traveling 670 million miles per hour relative to the spaceship. Is the light going 1,035 million miles per hour relative to Earth? Einstein says no! The light is moving at exactly the same speed relative to both the spaceship and Earth, even though the spaceship is moving at half the speed of light, and in the same direction as the light, relative to Earth.

To understand how light travels at the same speed relative to all observers, we have to rethink time and space. In fact, although all observers must agree on the speed of

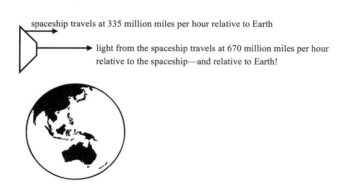

spaceship travels at 335 million miles per hour relative to Earth

light from the spaceship travels at 670 million miles per hour relative to the spaceship—and relative to Earth!

Figure 15 The spaceship travels at 335 million miles relative to Earth. The light from the spaceship's headlights travels at 670 million miles relative to the spaceship, and at exactly the same speed relative to Earth. *Source:* the image of Earth comes from publicdomainvectors.org

light in empty space, they disagree about other basic measurements of time and space. If one observer is moving at high speeds relative to another observer, the observers will disagree over the results from three types of basic observations: the time interval between two events, the lengths of objects, and even the order in which certain events occur.[1] Both observers are correct! Neither one is under some sort of delusion. The speed of light is rigid, so time and space must be fluid.

One of Einstein's astonishing truths is that the length of a car, for example, depends on who's measuring it. Specifically, the length of the car depends on how fast it's moving relative to the person who's measuring it. So if we ask, "What's the length of that car?" the question is ambiguous and incomplete, much like the question, "What's the distance to Atlanta?" We might want to know the distance to Atlanta from wherever we are when we ask the question ... or we might want to know the distance to Atlanta from Alpha Centauri. Analogously, we might want to know the length of the car when we're sitting in the car ... or we might want to know the length of the car when it zooms by at half the speed of light. *The car's length is not a property of the car alone*; the car's length depends also on the relative speed of the observer.

These relativistic effects are significant only at relativistic speeds, meaning speeds close to the speed of light. In daily life, the speed of one person relative to any other

The car's length is not a property of the car alone; the car's length depends also on the relative speed of the observer.

person is much less than the speed of light. Therefore, in daily life, we all agree on time intervals, lengths, and the order in which events occur. So relativity, like quantum physics, is part of the mysterious structure of the universe revealed only when technology enhances our awareness beyond everyday perception. Let's see how Einstein's mysterious truths arise from the single innocent fact that everyone agrees on: the speed of light in empty space.

Time Dilation

There's a standard way to prove that observers disagree about the time interval between two events. Imagine a room with a laser on the floor, aimed at a mirror on the ceiling. Consider the time interval between these two events: the emission of a brief pulse of light from the laser, and the arrival of the light back at its source after reflecting off the mirrored ceiling.

Distance = Speed × Time,

so the time interval between the emission of the light pulse and the return of the light pulse is

Time = Distance/Speed,

where Distance is 2H, twice the height of the room (figure 16), and Speed is the speed of light. An observer in the

H is the height of the room

Figure 16   The total distance traveled by the light pulse is 2H, twice the height of the room.

room thus finds that the time interval is 2H divided by the speed of light.

But now imagine that the room is actually a train car, and the train is traveling at a relativistic speed, v, relative to the ground. The train car is made out of a transparent material, so someone outside the train can observe the two events. Imagine someone standing on the ground outside the train, and let's call this person Grover (of the ground). Let's also imagine someone named Tracy (of the train), standing in the train car.

Grover witnesses the same two events: the emission of the light pulse from the laser on the floor of the train car, and the return of the light to its source. These two events occur at different locations relative to the ground because the train moves as the light travels. In fact, to Grover, the light must travel diagonally.

As shown in figure 17, the train is in one position when the light is emitted, another position when the light reflects off the ceiling, and a third position when the light returns to its source. The total length of the two

Figure 17 Three positions of the train car, shown as dashed rectangles, relative to the ground: The light is emitted in the first position. The light reflects off the ceiling in the second position. The light returns to its source in the third position. The diagonal arrows represent the path of the light, and the horizontal arrow represents the motion of the train car.

diagonal arrows is the distance traveled by the light, as seen by Grover. So the distance traveled by light, between its emission and return to source, depends on who's watching.

Both observers agree that Time = Distance/Speed. And both observers agree that Speed is the speed of light, which is the same for everyone. But the distance is not the same. As seen by Grover, light travels a greater distance than 2H. Therefore, Grover measures a larger time interval between the two events than the train observer does. Science-fiction writers love this fact: more time elapses for people on Earth than for high-speed space travelers. This is not a technological failure of clocks, or a biological response to space travel, or any kind of flawed perspective. Time itself elapses differently for people traveling at different speeds.

Time dilation is often summarized as "moving clocks run slow." Tracy uses a clock affixed to the floor of the train car to measure the time interval between the two events occurring there (the emission and return of the light pulse). Grover disagrees with the time interval shown by Tracy's clock. Grover measures a larger time interval; therefore, Grover complains that Tracy's moving clock runs slow.

It may be confusing to recognize that from Tracy's point of view, it's Grover who is moving. Therefore, according to Tracy, it's Grover's clock that runs slow! In fact, each observer complains that the other observer's clock runs slow. We will not resolve this paradox in detail. But the resolution lies in the two other basic results of relativity: length contraction, and disagreement about chronological order. We turn to these shortly.

We may be curious about a related conundrum, the twin paradox. Suppose two twins are born on Earth. One twin hops aboard a spaceship, travels at relativistic speed, and returns to Earth to find that her twin has aged much more than she has. In fact, both twins agree on this fact when they reunite on Earth. On the one hand, this is a simple illustration of time dilation: the traveling twin's clocks (including her physiological clock) ran slow. On the other hand, from the traveling twin's point of view, the Earth accelerated away and then returned, while the spaceship remained still. Isn't this a valid perspective? From a perspective at rest in the spaceship, the earthbound twin was the high-speed traveler who should have aged less.

The resolution of this paradox is that our intuition is correct: it really is the spaceship that accelerates away from Earth. If you stand on an accelerating bus, you feel the lurch that may cause you to lose your balance. The people standing on the sidewalk don't feel the lurch of acceleration, even though, from a perspective at rest in the bus, it's the sidewalk that accelerates away. Using accelerometers, we can prove that the bus accelerates relative to Earth; and Earth remains at rest.[2] Similarly, the traveling twin experiences acceleration, but the earthbound twin does not.

There's really quite a lot going on in the twin paradox, but we can identify the basic stages:

1. The twins start out on Earth. They're the same age.

2. One twin hops aboard the spaceship and accelerates away from Earth. The traveling twin is accelerating, and the earthbound twin is not. The traveling twin can confirm this fact by observing an accelerometer on the spaceship.

3. The traveling twin then travels at constant, relativistic speed through space. There's no acceleration now (the spaceship is coasting), so each twin's perspective is valid: each twin sees that the other's clocks are running slow. Both twins are correct; time elapses differently for different observers.

4. The spaceship decelerates to rest, and then accelerates back toward Earth for the voyage home. The earthbound twin, however, experiences neither acceleration nor deceleration.

5. The traveling twin again travels at constant, relativistic speed, this time toward Earth. Since there's no acceleration, each twin again makes the valid observation that the other twin's clocks run slow.

6. The traveling twin decelerates to land on Earth, while the earthbound twin remains at rest.

7. Both twins are at rest on Earth. The twin that traveled is younger because only that twin experienced acceleration and deceleration.

The bottom line is that the laws of physics are the same for all observers traveling at a constant speed (no acceleration). To make sense of what's happening to someone who accelerates up to relativistic speeds, we can take the point of view of someone at rest (or moving at constant speed) the whole time.[3]

Length Contraction

Recall the experiment used to prove time dilation. Suppose that the laser happens to be directly over a railroad tie when it emits its pulse of light. Suppose too that the laser is directly over another railroad tie when the light returns to its source. Tracy and Grover now want to determine the distance between the two railroad ties. Both observers agree on two quantities: the speed of light, and the speed v at which they're moving relative to each other.

Since Distance = Speed × Time, both observers can use this formula to find the distance between the two railroad ties. For Tracy, Speed = v, the speed at which the railroad ties are whizzing past below her. Time is the time interval between the emission of the light (when the first railroad tie is directly below) and the return of the light to its source (when the second railroad tie is directly below).

For Grover, Speed is also v, the speed at which the train is racing by. Time is also the time interval between the emission of light (when the laser is directly above the first railroad tie) and the return of the light to its source (when the laser is directly above the second railroad tie). But we found above that the two observers disagree about this time interval! Grover measures a greater time interval. Since distance is proportional to time, Grover measures a greater distance between the two railroad ties than Tracy does.

Tracy measures a shorter distance between the railroad ties. Relative to Tracy, the train tracks are moving, and the distance along the tracks is contracted, compared with Grover's measurements. This is the contraction of the length of moving objects, along the direction of motion. Both observers agree on the height of the train because length contraction does not affect distances perpendicular to the direction of motion.

We saw that each observer complains that the other observer's clocks run slow. Similarly, each observer complains that the other observer's lengths are contracted (but only along the direction of motion). Relative to Grover, the train is moving, so Grover measures that the train has a shorter length than is measured by Tracy. How can each observer claim that the other observer's lengths are contracted? We will be able to resolve this paradox after establishing one more consequence of light's constant speed.

The Chronological Order of Events May Depend on Who's Observing Them

The third consequence of light's invariant speed may be the strangest. It's also the most significant for our discussion of quantum entanglement. Consider now a flash of light from the center of the train car. As seen by Tracy, the light simultaneously reaches the front wall and back wall of the train car (figure 18). The two events (the arrival of light at the front wall and the arrival of light at the back wall) are simultaneous. But only according to Tracy!

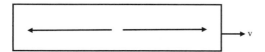

Figure 18 A flash of light from the center of the train car reaches both walls simultaneously—but only according to the observer on the train.

Grover sees the front wall traveling away from the light source. Similarly, the back wall is traveling toward the light source. Therefore, the light reaches the back wall first. Later, the light reaches the front wall. The two events, which were simultaneous in Tracy's eyes, occur in sequence according to Grover. The simultaneity of two events is not a universal fact but depends on who's observing.

We can imagine that Tracy has a stopwatch affixed to both walls of the train, and the stopwatches are designed to start counting when the pulse of light reaches them. Tracy sees both stopwatches start at the same time, so the two stopwatches are synchronized. Grover, on the other hand, sees the stopwatch on the back wall start first, so the stopwatch on the back wall shows a time in advance of the stopwatch on the front wall.

This discrepancy applies to all of Tracy's clocks, not just stopwatches. It's convenient to imagine that Tracy has lots of synchronized clocks along the length of the train. The further a clock toward the back of the train (to the left in figure 18), the further ahead the time that it shows, according to Grover.

This disagreement about chronology allows us to resolve the paradox about length contraction. Tracy sees that Grover's lengths are contracted along the direction of motion, and Grover sees that Tracy's lengths are contracted along the direction of motion. We can imagine that each observer has yardsticks aligned with the direction of motion. Each observer sees that the other observer's yardsticks are too short—less than a yard.

Grover wants to understand how Tracy, with her contracted yardsticks, can possibly claim that Grover's yardsticks are too short—shorter than hers. Grover sees clearly that Tracy's yardsticks are too short—shorter than his. Grover recognizes that the length of a yardstick is the distance between the two endpoints of the yardstick. So, to measure the length of a yardstick that's moving past you, you can record the positions of the two endpoints—but you have to be careful to record those two positions at the same time! If an arrow's flying by you, and you record the position of front tip before you record the position of the back tip, the distance between the two positions is less than the length of the arrow.

Grover sees that Tracy's clocks not only run slow; they disagree with one another. So when Tracy records the positions of the endpoints of Grover's yardstick, she records the two positions at the same time—according to her own clocks. But Grover sees that Tracy's clock toward the left shows a later time than her clock toward the right.

So Grover sees that everything that happens to the left occurs too soon; when Tracy tries to perform two simultaneous measurements, she actually first performs the measurement on the left, followed by the measurement on the right. (Tracy herself observes that she performs the measurements at the same time.)

This resolves the paradox, because Grover sees the following: Making a mark on the transparent floor of her train car, Tracy identifies the point directly above the left endpoint of Grover's yardstick. (To Tracy, the left endpoint is like the front tip of a moving arrow.) According to Grover, time elapses before Tracy marks the point directly above the right endpoint of Grover's yardstick (which, to Tracy, is like the trailing end of the moving arrow). Thus the distance between the two marked points, according to Grover, is unfairly reduced; the two endpoints were recorded at different times. During this time interval, the train has moved relative to the yardstick, so of course the recorded distance between the two positions differs from the true length. Tracy finds that the distance between the endpoints is less than a meter—even using her own contracted (as Grover sees it) yardstick.

Tracy's perspective is equally valid. She can explain why Grover, even with his shrunken yardsticks, claims that her yardsticks are the shrunken ones. She recognizes that Grover thinks that he's simultaneously recording the positions of the endpoints of her yardstick, but he actually allows time to pass between the two measurements.

Again, each observer is making a mistake only from the perspective of the other observer. Both observers make equally legitimate claims about time intervals, lengths, and the order in which events occur. These properties of space and time are not universal but depend on who's observing them.

Now consider an observer traveling to the right even more quickly than the train. To this observer, the train is receding to the left. Thus, to the speedy observer, the left wall of the train is receding from the light source, and the right wall is moving toward it. The speedy observer sees the light arrive at the right wall before it arrives at the left wall—opposite to the chronological order observed by Grover!

The length-contraction paradox was resolved, but at what price? Have new paradoxes sprung up? If the order of events depends on who's looking, we seem to have the possibility of time travel and its attendant paradoxes. For example, consider the following two events:

1. I trip at the top of the stairs.

2. I land on my face at the bottom of the stairs.

According to my observations, Event 1 occurs before Event 2. But if you're traveling at relativistic speed, might you see Event 2 occur first? If you see me land on my face before I even trip, could you stop Event 1 from happening, perhaps by sealing off the top of the stairway? But if you prevent Event 1 from happening, how were you able to witness Event 2, which was caused by Event 1?

Einstein showed that these paradoxes are avoided, as long as nothing travels faster than light. If nothing travels faster than light, then all observers agree that a cause occurs before all of its effects; an effect occurs after all of its causes. The order of events is different for different observers only if the events are causally unconnected.

But what of that key disclaimer, "as long as nothing travels faster than light"? Does quantum entanglement violate this condition?

The Apparent Conflict with Quantum Entanglement

Let's return now to quantum entanglement and the observation of two entangled photons. If the measurement of one photon physically alters the other, the effect occurs instantly and with undiminished impact over any distance. On the other hand, Einstein's relativity implies that nothing travels faster than light, whose speed is 670 million miles per hour. Suppose that I'm 670 million miles away from you. Halfway between us, a pair of entangled photons is emitted. These are the entangled photons we met in chapter 3: they are just as likely to be horizontally polarized as vertically polarized. If both photons encounter a horizontal polarizer, they will either both pass through or both be blocked.

Suppose, too, that we each decide to place a horizontal polarizer in the path of the photon heading our way. You find that your photon passes through the polarizer, and you want to convey this information to me in the fastest possible way: via radio waves, which travel at the speed of light. The radio waves hurtle through space for a full hour before they reach me. By the time your radio message arrives, it's old news. For a whole hour, I've already known the result of your measurement because I too saw my photon pass through a horizontal polarizer. Have we found a way to beat relativity and communicate faster than the speed of light?

Actually, the measurement of one photon cannot in any way be used to send messages via its distant twin. Both experimenters can agree to set their polarizers to the same angle, but the first photon measured is just as likely to be transmitted as blocked. The distant photon must do the same thing, but this doesn't allow us to encode any message since the "signal" we're sending is a random outcome over which we have no control. We can reconcile relativity with quantum entanglement by clarifying that neither mass nor messages can travel faster than light. The subtle linkage between entangled particles conveys neither mass nor messages.

Spooky action at a distance is consistent with even the strangest consequence of relativity: the chronological order of events may depend on who's observing. You and I may agree that *you* measure your photon before I measure mine, but someone traveling exceptionally fast may observe, instead, that *my* measurement occurs first. The speedy observer sees the measurements occur in the opposite order, but the result is the same: the outcome of the first measurement is random, and the outcome of the second measurement is compelled to be the same.

DIRECT OBSERVATION IS THE ONLY REALITY?

Experiments contradict local realism, but it's easier to reject falsehood than to establish the truth. The truth is elusive. We cannot observe photons prior to observation—so we do not have direct evidence that the observation of one photon affects another. Debate over these points has resulted in a host of interpretations of quantum mechanics.

Let's review, once more, the creation of two entangled photons that are just as likely to be horizontally polarized as vertically polarized. Let's place a horizontal polarizer in front of each photon. Half the time, both photons pass through the polarizers. Half the time, neither photon passes through.

Let's name the photons A and B. Suppose Photon A reaches its polarizer before Photon B. If Photon A passes through the polarizer, then we know that Photon B is certain to pass through its horizontal polarizer when it gets there. But what really changed when Photon A passed through its polarizer? Did Photon A change? Did both photons change? Did neither change? Or do the changes take place only after the photons reach detectors, or after the detectors communicate the result to a circuit board?

If I claim that the size of my right foot changes when I measure my left foot, we would expect to observe this directly: when I hold a ruler up to my left foot, we should be able to watch my right foot shrink or expand, or perhaps transform from fuzziness to solidity. Similarly, we want to observe Photon B, both before and after Photon A is measured, to see if anything changes. But then, the first observation of Photon B would be a measurement, which may affect the state of Photon A!

My claim is that both photons are transformed by the first observation of either photon. Thus this transformation can never be observed; we can't perform any observation prior to the first observation. So we can never watch one particle change in response to the measurement of its twin. The innermost workings of nature remain forever out of reach. The quest for complete understanding is always an unscratchable itch. The only fact that's (almost) certain is that local realism cannot account for measured results.

Local realism is defeated by violations of Bell inequalities, which is why local realism is the negative space of quantum physics: local realism is the excluded explanation. If we reject local realism, what's left? Are the only remaining views of reality mystical? Does quantum mechanics, after all, say something mystical about the universe? We can no longer argue that physics is merely a set of formulas for predicting experimental outcomes, disjoint from philosophical considerations: Bell inequalities show that experiment has overruled a plausible philosophical assumption. There are many alternative assumptions, but none are especially plausible, and all have their partisans.

Indeed, there are many philosophical interpretations of quantum mechanics. I will not try to compile a complete list, or give equal attention to leading viewpoints, or even to classify the viewpoints in a standard way. But I will consider four categories of responses to Bell inequalities:

1. Ferret out assumptions we didn't know we were making.

2. Abandon both locality and realism.

3. Abandon locality to save realism.

4. Abandon realism to save locality.

We can no longer argue that physics is merely a set of formulas for predicting experimental outcomes, disjoint from philosophical considerations: Bell inequalities show that experiment has overruled a plausible philosophical assumption.

Ferret Out Assumptions We Didn't Know We Were Making

If we hope to cling to local realism, like a life preserver in a stormy sea, we need to identify another assumption that may be false. We then blame this other assumption for the incompatibility between experiment and the mathematical constraints that we derived. If this other assumption is to blame, then local realism may be innocent.

First, let's think about an assumption that seems identical to realism: the unstated assumption of *counterfactual definiteness*.[1] This is the assumption that even though each photon goes through a polarizer set to a single angle, we can specify what the photon *would* do if it went through a polarizer at a *different* angle. The assumption of realism—that the photon has properties that predetermine its response to any chosen polarizer angle—seems to *require* counterfactual definiteness. In a moment, we'll distinguish realism from counterfactual definiteness.

In normal life, counterfactual definiteness seems reasonable. For example, I'm not jumping right now, so it's *counterfactual* to discuss what would *definitely* happen if I were to jump. Yet I can say with confidence that if I jump, I will come back down. And if I drop my pen, it will fall. And if I clap, I will hear the sound. It seems obvious that all these statements are true—and that's because counterfactual definiteness is so innocuous, we assume it all the time.

We already know that quantum particles defy our expectations in many ways. We might as well ask whether there's something fundamentally forbidden about specifying what a particle *would* do in any situation other than the one it actually experiences. If we reject counterfactual definiteness, can we save realism?

The distinction between realism and counterfactual definiteness becomes clearer if we consider the viewpoint of **superdeterminism**.[2] According to superdeterminism, there's no free will. The entire universe is a Rube Goldberg device evolving inexorably along its predetermined course. Every future occurrence, down to the minutest detail, was predetermined at the moment of the Big Bang. Free will is an illusion, and if we believe in this illusion, it's only because we were predestined to do so.

All of the Bell inequalities are derived under the assumption that the experimenters can freely choose polarizer angles, such as 0°, 30°, or 60°. Since the photons don't "know" the angles that they'll encounter, they have to be prepared (they have to have hidden properties) for all possible angles. In a superdetermined universe, the photons have a lot less preparation to do. Each photon needs a property only for the *single* polarizer angle that it's certain to encounter. The assumption of realism is thus valid; the photon has its *single* property (causing it to be transmitted or blocked) all along, even before we measure it. Although this explanation preserves realism, it's still very strange.

Somehow each photon "knows" exactly which polarizer angle it will encounter. The two photons no longer have to collude with each other, but instead each has to collude with its own polarizer before it even gets there. We can preserve locality by arguing that the collusion took place during the Big Bang, when everything was scrunched into one locale.

In any case, counterfactual definiteness isn't valid in a superdetermined world. We can't specify what a photon would do in any measurement except the one it actually experiences because there's never any possibility of anything else.

Another viewpoint that undermines counterfactual definiteness is the **many-worlds interpretation** of quantum mechanics.[3] In this view, all possible outcomes of a measurement are real—in parallel universes! When the measurement is performed, the world splits—the photons are vertically polarized in one world, and horizontally polarized in the other. (I believe adherents of this interpretation prefer different terminology: The only reality is the sum of all possible outcomes. So reality itself isn't splitting; there are just new branches within the single reality, and we're conscious of only one of the branches.)

As unrealistic as it seems, the many-worlds interpretation is based on realism. Quantum mechanics represents the state of our entangled photons as a sum of two mutually exclusive outcomes, horizontal polarization and vertical polarization. This sum is considered the ultimate reality in the many-worlds interpretation. The measurement causes the terms in the sum to split off into separate worlds. The sum of all the worlds remains the single deep reality, but we perceive only the one world we inhabit. In a sense, when I measure a photon's polarization, I'm not changing the photon, which always existed as a sum of vertical and horizontal polarization; I'm changing *myself*, splitting into someone who observes vertical polarization, and someone who observes horizontal polarization.

Now we want to explore whether counterfactual definiteness makes sense in the many-worlds interpretation. In a particular world, can we specify what a photon would do if the polarizer were set differently from how it's *actually* set? Well, what would make the experimenter decide to set it differently? Is it random? Let's imagine that a random quantum event determines the direction of the polarizer. For example, we might set up a light source to emit a single photon. Let's call this photon the Decider. We arrange an experiment so that the Decider has a 50 percent chance of being vertically polarized, and a 50 percent chance of being horizontally polarized. Suppose the polarization of *this* photon sets the angle of one of the polarizers in our entanglement experiment.

But wait—the world split when the Decider photon was measured! We can't talk about what the entangled photon would do if the polarizer were set differently because *that* occurs only in a different universe!

When describing the many-worlds interpretation, authors like to give the disclaimer, "This sounds like science fiction." And yet a surprising number of physicists actually believe in it. It's amazing that the same profession that gives us airplanes and computer chips also tells us that our whole entire *universe* may be a vanishingly tiny speck in an exploding infinity of parallel worlds.

Indeed, the many-worlds interpretation has its partisans because of how it resolves the *measurement problem* in quantum mechanics.[4] The measurement problem is not specific to entangled particles. Even a single particle is in a fundamentally undecided, unknowable state prior to measurement. The measurement forces the particle to settle into a state with a more exact value of one property (its position, for example), while another property (its speed) unavoidably becomes more uncertain. Similarly, when measuring a photon's polarization in a horizontal or vertical direction, we learn whether it's horizontally or vertically polarized, but we lose any information we may have had about whether it was polarized in the 45° or –45° direction. The loss of information about one property, when measuring a different property, is Heisenberg's famous uncertainty principle.

But at exactly what point does the measurement take place? This is a key question, since such a fundamental

The same profession that gives us airplanes and computer chips also tells us that our whole entire *universe* may be a vanishingly tiny speck in an exploding infinity of parallel worlds.

transformation takes place. The particle seemingly transmutes into something it wasn't just before. Does the transmutation occur when the photon passes through the polarizer? Or when it reaches the detector? Or when the detector sends an electronic signal to a circuit board? Or when the circuit board transmits the message to a computer? Or when the computer displays 0 or 1? Or when a conscious observer sees the result on the computer screen? Some physicists have actually proposed that *consciousness creates objectively real states*. Before registering in someone's consciousness, the photon is in a fundamentally undetermined and unknowable state—and so is everything it encounters on its way, in an avalanche of indeterminacy! In this view, the computer screen is in some unimaginable combination of showing *both mutually exclusive outcomes* before the conscious observer comes along.

Measurement is a problem because a measurement disrupts the smooth evolution of a quantum state. The fundamental equation of quantum physics does one thing very well: it specifies the probabilities of measurable outcomes, and how these probabilities change over time. As soon as a measurement is made, all the outcomes that weren't measured get thrown out of the equation. This "throwing out" process is external to the equation and no one fully understands it—and it doesn't happen at all in the many-worlds interpretation.[5] In the many-worlds interpretation, no outcomes get thrown out because all possible outcomes coexist in parallel worlds.

Now we'll think about a few more assumptions underlying the derivations of the Bell inequalities. There's an assumption called *fair sampling*: no detectors are 100 percent efficient; the detectors miss a large fraction of the photons that reach them. For example, suppose a detector responds to 20 percent of the photons that reach it. Our unstated assumption is then that each photon arriving at the detector has the same 20 percent chance of detection; the system is not somehow rigged. This fair-sampling assumption is also called the *detection loophole*. If we reject the fair-sampling assumption, we could make the (outrageous?) claim that the detector somehow favors photons that violate Bell inequalities, just to fool us; only the *detected* photons violate Bell inequalities, and if we replaced our detectors with ideal (100 percent efficient) detectors, the Bell inequalities would in fact be satisfied.

We've also assumed that the photons don't "know" in advance the angle of the polarizer they'll encounter; this is why realism requires the photons to have preset outcomes for all possible polarizer angles. What if the photons can somehow sense in advance the angles of the polarizers? Then we can't derive any Bell inequalities because the photons need to have just a single preset outcome. (We followed this line of reasoning in the discussion of superdeterminism.) We're now imagining some form of signal from the polarizers to the photons, so that the photons are alerted to exactly which polarizer angles they'll encounter.

To me, it's much *less* spooky to think that the measurement of one photon affects the other, than to think that the photons somehow sense the angles of polarizers they haven't arrived at yet. If we imagine that the photons may somehow receive information from the polarizers before they get there, we have what's called the *locality loophole*. The idea is that if information travels (no faster than light) from the polarizer to the photons, then this information is available locally, at the photons' original position. This is contrasted with the idea that the measurement of one photon instantly (faster than the speed of light) affects the other (nonlocal) photon.

From 1982 until 2015, various experiments closed either the detection loophole or the locality loophole. For example, in 1982, Alain Aspect led an experiment that effectively rotated the polarizers as the photons were traveling to them, so the photons couldn't have any advance notice as to which polarizer angle they'd encounter.[6] Other experiments used detectors with high efficiency to close the detection loophole. But prior to 2015, a true zealot could still insist that local realism was *not* to blame for the experimental violation of mathematical constraints. Finally, in 2015, both loopholes were closed simultaneously in a single experiment.[7]

To close the locality loophole, the polarizer angles have to be chosen unpredictably so that the photons can't have any advance notice of what they'll encounter. We can place a random number generator at each polarizer to choose the angle. But what if some unknown, common cause affects both the random number generators and the photons? Then, the photons *could* have predetermined properties all along, while still violating Bell inequalities, due to the unknown influence tampering with the random number generators. This loophole is called the *freedom-of-choice loophole*. It challenges our assumption that the choice of polarizer angles can be made freely, independent of the properties of incoming photons. Theoretical work has shown that local realism can be preserved if the tampering influence is minimal; we don't need to go to the extreme of superdeterminism.[8]

How can we close the freedom-of-choice loophole? We need to rule out a tampering influence, which travels no faster than the speed of light. Some physicists have used light from distant stars to set the angle of polarizers, and the results were the same as always: Bell inequalities were violated.[9] The starlight was emitted hundreds of years ago, and we assume that the stellar photons were unaltered during their long journey to Earth. If a tampering influence exists, it must have planned ahead by hundreds of years, before the starlight was emitted, just to produce a Bell inequality violation. This hypothetical, tampering influence is like a patient villain with an extremely perplexing goal.

In another experiment to close the freedom-of-choice loophole, about 100,000 people from around the world generated random numbers.[10] The random numbers were used to set the polarizer angles (or equivalent analyzer settings) in tests of Bell inequalities. Participants generated random numbers by playing a video game online.[11] The Bell inequalities were violated, as usual. We conclude that local realism was defeated: the entangled particles did not have definite properties prior to measurement, or if they did, the measurement of one particle affected the other. Alternatively, a superdeterministic power governed the seemingly random choices of 100,000 people so that their choices corresponded with properties that the entangled particles had prior to measurement. In either case, common sense cannot account for the results.

Let's consider a final assumption that we've made all along, which also seems like common sense: the two entangled particles, when separated by an arbitrarily large distance, are in two different places, not a single place. How could this possibly be untrue? Well, what if the two entangled particles are connected by a wormhole, which is a shortcut through space and time (like Madeleine L'Engle's "wrinkle in time")? Since both ends of a wormhole are actually the same point, then no matter how far apart the entangled particles are, they occupy the same position!

Leonard Susskind and Juan Maldacena advanced this idea several years ago, in 2013.[12] The succinct nickname for this conjecture is ER = EPR.

"EPR" refers to the paper Einstein coauthored with Boris Podolsky and Nathan Rosen in 1935, arguing that entanglement reveals a flaw in quantum mechanics: a more complete theory is needed to specify the exact outcome of any possible measurement. Less than two months after the EPR paper was published, Einstein and Rosen (ER) published a paper about (what we now call) wormholes.[13] If ER=EPR, then entangled particles are even stranger than we thought, connected via invisible tunnels through space and time!

In fact, a favored view among physicists is that **reality is a higher-dimensional space**.[14] Our ordinary ideas of space and time are inadequate to understand entanglement. To recognize our cognitive limitations, we can imagine a world with fewer dimensions than ours: imagine a society constrained to exist in a flat, geometric plane.[15] The two-dimensional people in this world have no concept of three-dimensional space because they have never experienced it.

Now imagine that a three-dimensional titan starts poking the tips of a fork through the two-dimensional world. The fork is poked at random moments through random locations. The two-dimensional people (quivering in terror) perceive the tines of the fork as four isolated, round blobs. They see no possible physical connections among the four blobs; they can completely encircle each blob with a string to prove that it's isolated from the others. The four blobs always appear at almost the same time, however, and they disappear at almost the same time. Although the two-dimensional scientists can't predict where or when the blobs will appear, the distance between adjacent blobs is always the same. (Perhaps the blobs expand slightly after they appear, and they shrink before they vanish, but the distance between the centers of the blobs is always the same.)

The two-dimensional scientists wonder if the appearance of one blob causes the three other blobs to appear, some distance away. Is this spooky action at a distance? The two-dimensional scientists are scratching their two-dimensional heads. Eventually, an idea forms in their two-dimensional brains. Perhaps the isolation of the blobs is an illusion; perhaps, in an unimaginable higher-dimensional space, the four blobs are part of a unified whole. The properties of one blob don't influence the properties of any other blob. Instead, the relationships among the blobs exist all along in a higher-dimensional space, which only occasionally intersects the familiar, two-dimensional reality.

This is how some physicists explain entanglement: We live in a cross-section of a higher-dimensional reality. Much like the two-dimensional scientists, we cannot intuitively understand causality in the higher dimension. Nicolas Gisin writes, "In a certain sense then, reality is something that happens in another space than our own, and what we perceive of it are just shadows, rather as in Plato's cave analogy used centuries ago to explain the difficulty in knowing the 'true reality.'"[16] This is an extraordinary statement. Scientists are stereotyped to equate reality with empirical data, but evidently some scientists equate reality with an invisible higher realm.

Let's review, once more, the assumptions of locality and realism. We've seen that the assumption of local realism imposes constraints on measurable quantities, and measurement violates those constraints. Thus, unless we embrace an exotic viewpoint like superdeterminism, we must abandon locality, realism, or both. As we've seen, locality and realism are both very reasonable:

- Locality means that the measurement of one object can't affect a distant object.

- Realism means that objects have properties that exist prior to measurement; measurement merely reveals properties that the objects had all along.

If physics forces us to abandon at least one of these common-sense notions, what's left, other than a higher-dimensional reality? Here are some possibilities.

Scientists are stereo-
typed to equate reality
with empirical data,
but evidently some
scientists equate reality
with an invisible higher
realm.

Abandon Both Locality and Realism

As I mentioned earlier, I find it expedient and straightforward to suppose that **measurement creates objectively real states**. If a photon passes through one vertical polarizer, it will pass through any number of vertical polarizers lined up in a row. The vertical polarization of the photon seems to be an objective fact, once it's measured in the first place.

A pair of entangled photons shares a single state, and the initial state is noncommittal. The photons do not have properties that predetermine their behavior at polarizers; they're not predestined to pass through polarizers or get blocked by polarizers. Let's suppose both polarizers are set to the same angle. If we first measure one photon in the pair, the measurement instantly creates a definite state for *both* photons. If we believe that the state is objectively real, then the measurement of one photon physically alters the distant photon. This is the spooky action at a distance that Einstein decried (and that many physicists continue to reject). We can't prove that the measurement of one photon changes the other because we can't watch the change take place; we can't make an observation prior to the first observation on either photon. But for exactly this reason, we can't prove that the measurement of one photon *doesn't* affect the other. For all practical purposes—predictions of outcomes—the measurement of one photon indeed affects both. Both photons transform from a noncommittal state to a known state. As Tim Maudlin writes, "Going from *not having a physical state* to *having a physical state* is some sort of change, call it what you will!"[17]

Even if spooky action at a distance is real, is it any spookier than other influences, like gravity? If you lived a thousand years ago and someone told you, "The only thing stopping Earth from floating off into the endless midnight of black space is a wrenching attractive force from the distant sun," wouldn't you think that was spooky? The only reason gravity doesn't seem spooky is that we're so familiar with it; we've absorbed it into our intuition about how the world works.

To me, spooky action at a distance is disappointingly unalarming. The real mystery resides in the measurement problem: Exactly what were the photons like before they were measured? How does a measurement transform their polarization from something essentially undefinable to something definite? And when does this transformation take place: What physical process functions as a measurement? And why don't we observe quantum effects on a large scale? Why can't we be in two places at once, or dead and alive at the same time?

Some physicists believe these questions have been partially answered by *quantum decoherence*: when objects are jostled by the surrounding air molecules and photons, they lose the ability to be in two mutually exclusive states at the same time. One state survives, and the other state dissipates. This process is explained through quantum equations. What's not explained is how the surviving state is chosen: the selection of the state remains a purely random process.

Can We Save Realism by Rejecting Locality?

There have been attempts to formulate nonlocal, realistic theories. David Bohm's theory is the most famous example. If we accept nonlocality, we have a chance of preserving realism. We've assumed all along that each photon is unaffected by the other photon's polarizer. Why, indeed, would a photon be affected by a distant sheet of plastic that it never even approaches? If your polarizer affects my photon, it must be indirect, through your polarizer's effect on your photon (which is entangled with mine).

In 2003, Anthony Leggett derived a generalized Bell inequality. We recall that ordinary Bell inequalities are based on two assumptions: realism and locality. Leggett retained the assumption of realism, but permitted a restricted form of nonlocality.[18] Quantum mechanics and measurements violate Leggett's inequality, but physicists dispute the significance of this result.[19]

Can We Save Locality By Rejecting Realism?

We need to identify just a single false assumption to explain why Bell's constraints do not apply to real particles. Could the false assumption be realism, such that locality may be valid? If we reject realism, then particles are not predestined to behave any particular way when they are eventually measured. Thus one photon in an entangled pair is not predestined to be vertically polarized, even if that's what the measurement ultimately shows. But when both polarizers are vertical, the two photons always do the same thing: they both pass through, or they're both blocked. If the photons are in a fundamentally undecided state before measurement, how can they possibly arrange to always behave identically when the polarizer angles are identical? Nonlocality is much more obviously *necessary* when realism is rejected. Indeed, Einstein's objection to quantum mechanics was that its lack of realism necessitates nonlocality.

But we can insist (boldly? shrilly? petulantly?) that **direct observation is the only scientific reality**. This assertion has been made in a variety of forms, starting with Niels Bohr's Copenhagen interpretation. An extreme version of this idea is called **genuine fortuitousness**, which denies the existence of microscopic particles! In this view, there are probabilities of responses from our detectors, but we shouldn't say that the detectors are actually detecting anything.[20]

"Direct observation is the only scientific reality" takes a less extreme, though still brazen, form, in a recent interpretation of quantum mechanics called **QBism** (pronounced "cubism" to deliberately create a sense of radical departure from established norms).[21] QBism is the abbreviation of "quantum Bayesianism." In Bayesian statistics, probabilities are updated as new information comes in.

For example, one time I was visiting Chicago. Coming out of a train I noticed that my wallet was gone. The departing train receded before my saddened eyes. I asked the transit staff if there was a lost and found. They gave me the phone number, but they told me that there was no point because I had been pickpocketed. I called the lost and found, but it was closed for the day. I spent the whole harrowing night believing that Chicago was a city of villains. Even if I got a new wallet, I would surely be robbed again.

The next morning, I called the lost and found, and learned that someone had turned in my wallet, with all $136 in it! I immediately reversed my judgment of Chicago. Chicago was a city of good Samaritans, and I could expect nothing but kindness from strangers.

The daily risk of crime in Chicago at no point actually changed over that 12-hour period. My subjective judgment of the risk, however, underwent two drastic updates.

In QBism, quantum mechanical probabilities are subjective judgments. There's no such thing as an absolutely accurate, objective probability "out there." Quantum mechanics is a tool for making our subjective judgments as accurate as possible. Different people will assign different probabilities to the same event if they have different information about it. Before a photon's polarization is measured, you and I may agree that the probability of vertical polarization is 50 percent. If you do the measurement and the photon is found to be vertically polarized, you update the probability to 100 percent. If I'm out of the room, I still think it's 50 percent until you give me the news. Until then, 50 percent and 100 percent are equally legitimate probabilities in the sense that they're both based on the best information available to the person.

Thus, according to QBists, there is *absolutely no action at a distance*. If I measure the polarization of one entangled photon and find it to be horizontally polarized, I immediately believe with 100 percent certainty that the other photon will also be horizontally polarized. If the other photon is traveling to you, a full light-year away, I have no way of communicating my knowledge to you before the other photon arrives; your photon had too much of a head start, even if I send you the news at the speed of light. For the whole year, you'll continue to believe that the probability of horizontal polarization is 50 percent, while I know it's really 100 percent. According to

QBism, we're both right! We're both applying quantum mechanics as accurately as possible with the information we have.

QBists refuse (humbly? peevishly?) to assign a cause to the observed correlations between entangled photons. The correlations are a fact of nature, and quantum mechanics gives us the math to accurately predict them. Any speculation as to how the correlations come about is outside the scope of physical science. (This approach is sometimes called "shut up and calculate.") Since QBist physicists don't speculate about underlying causality, the speculation and discussion must therefore come from ... philosophers ... or from theologians, poets, or science-fiction writers?

I don't permanently encamp with the QBists. But on occasion, QBism feels like an invigorating breeze that clears away a cloying miasma of confusion. QBism fends off the questions of what a particle's like before measurement, what constitutes a measurement, and what is the underlying deep reality. QBism ejects these questions from the realm of science because they all inquire about something that can never be scientifically determined: the state of an object before it's observed. It's not wrong to speculate about what a particle's like before it's measured, or to wonder what invisible *mechanism* enables one photon to always behave like its twin; it's just that we step outside of QBist science when we speculate about things that can never be directly observed.

What happens to objects that no one's looking at? Does the seemingly solid world dissolve into the phantasms and mirages of our own assumptions and mental images? The visible universe does not completely blink out of QBist existence when we close our eyes; the lapse in observation is filled in by the subjective judgment that the world is still there. QBism preserves our common sense. Quantum mechanics is classified as a prediction tool, not a gateway to ultimate reality.

QBism sweeps the cobwebby spookiness out of quantum physics (and into someone else's discipline). There's no action at a distance, and there's no speculation (within physics) about what particles are doing when we're not looking at them. But we can push this idea in a direction unintended by QBism's inventors. If we really believe that direct observation is the only reality, then, looking at the night sky is a single truth; observer and observed cannot be logically separated. And the quest to preserve locality leads to unification with everything we see.

GLOSSARY

Bell inequality
A constraint imposed on a measurable quantity by local realism, with the added assumption that the experimenter can freely choose detector settings.

Coincidence
In the context of quantum optics, simultaneous detection of two photons.

Copenhagen interpretation
A family of interpretations that reject realism. Quantum mechanics is a tool for predicting measurement outcomes, not a description of what particles are doing when we're not looking at them.

Counterfactual definiteness
The assumption that we can state what the result of a measurement would be under conditions different from actual conditions. Counterfactual definiteness is assumed implicitly in the derivation of Bell inequalities. Bell inequalities can be derived assuming that particles have properties that predetermine the results of all possible measurements.

Detection loophole
A consequence of limited efficiencies of particle detectors. Unless efficiency is sufficiently high, we must assume that all incoming particles have an equal probability of detection; the detector does not preferentially detect particles that violate Bell inequalities.

Freedom-of-choice loophole
A consideration of the possibility that particles somehow affect detector settings, or that an unknown influence affects both the particles and the detector settings.

Hidden variables
Unknown properties or influences that determine the outcomes of all possible measurements.

Length contraction
The shortening of lengths of objects moving at relativistic speeds.

Locality
The assumption that the measurement of a particle is unaffected by the measurement of a distant particle.

Locality loophole
A consequence of hypothetical communication between measuring equipment and entangled particles. The locality loophole is closed if the measuring equipment makes unpredictable changes so rapidly that no communication (up to the speed of light) can enable the entangled particles to violate a Bell inequality.

Local realism
The combination of two assumptions: objects have properties that exist regardless of whether anyone is observing them or knows what they are, and the measurement of one object is unaffected by the measurement of a distant object. Local realism is common sense, and yet it is rigorously contradicted by measurements of entangled particles.

Many-worlds interpretation
The view that the deep reality is the sum of all possible states of particles. Mutually exclusive results occur in parallel universes, which are branches of the one deep reality.

Photon
A particle of light.

Polarizer
A material that transmits light whose electric field is restricted to a single plane.

QBism
A relatively recent interpretation of quantum mechanics, which emphasizes that all probabilities are subjective judgments.

Realism

The assumption that objects have properties that exist regardless of whether anyone is observing them or knows what they are. Measurement reveals properties that objects already had.

Relativistic speed

A speed close to the speed of light.

Superdeterminism

The view that everything in the universe, down to the smallest detail, was predetermined from the moment of the Big Bang.

Time dilation

The slow elapsing of time of anything moving at relativistic speeds.

Wormhole

A hypothetical shortcut through space and time.

Preface

1. A. Einstein, M. Born, and H. Born (1971), *The Born-Einstein Letters* (Macmillan).

Introduction

2. G. Orwell (1949), *1984* (Secker & Warburg).

Chapter 1

1. This statement may be controversial, as we'll see in chapter 6. It's certainly true that measuring one particle helps us predict the outcome of measuring the other particle—with 100% certainty in some cases. It's also certainly true that *something* changes as a result of the measurement of the first particle; the measurement does not merely reveal properties that the particle had all along. Combining these facts, we are tempted to say that the measurement of one particle instantly affects both particles, but not all physicists would agree.

2. To be fair, we must recognize that geocentrism was supported by legitimate arguments based on observation. For example, we don't feel the motion of the earth, in contrast to the fact that we feel the motion of a ship. Also, with the unaided eye, we can't observe *stellar parallax*: apparent motion of stars relative to one another due to the motion of the earth around the sun.

3. For simplicity, I'm assuming that the hidden variable is deterministic (not random). The key feature of a hidden variable, however, is the realism: the hidden variable, *even if it is truly random*, determines the properties of an object regardless of whether the object is ever observed.

Chapter 2

1. This effect was first demonstrated by W. Gerlach and O. Stern (1922), "Der experimentelle Nachweis der Richtungsquantelung im Magnetfeld," *Zeitschrift für Physik* 9: 349–352.

2. A. Einstein, B. Podolsky, and N. Rosen (1935), "Can quantum-mechanical description of physical reality be considered complete?" *Physical Review* 47: 777–780.

3. J. Bell (1964), "On the Einstein Podolsky Rosen Paradox," *Physics* 1: 195–200. Curiously, the date of this paper is sometimes incorrectly given as 1965,

sometimes even in Bell's own book (2004), *Speakable and Unspeakable in Quantum Mechanics*, 2nd ed. (Cambridge University Press).

4. This average number is called the quantum correlation.

Chapter 3

1. This process is called spontaneous parametric downconversion.

2. This splitting respects conservation of energy: each infrared photon has half the energy of the original violet photon.

3. Even if the hidden variables theory is probabilistic, such that properties are randomly assigned to photons at the moment they're created, this theory at least conforms to local realism. Quantum theory, in contrast, leaves the properties of the photons undetermined until measurement.

4. S. J. Freedman and J. F. Clauser (1972), "Experimental test of local hidden-variable theories," *Physical Review Letters* 28: 938–941.

5. J. Brody and C. Selton (2018), "Quantum entanglement with Freedman's inequality," *American Journal of Physics* 86: 412–416.

6. Philip Ball (2018), *Beyond Weird: Why Everything You Thought You Knew about Quantum Physics Is Different* (Basic Books).

Chapter 4

1. This example is based partially on V. Scarani (2006), *Quantum Physics—A First Encounter: Interference, Entanglement, and Reality* (Oxford University Press). The original idea is from J. F. Clauser, M. A. Horne, A. Shimony, and R. A. Holt (1969), "Proposed experiment to test local hidden-variable theories," *Physical Review Letters* 23: 880–884.

2. A realistic analyzer is more complicated because we don't want to block any photons. For example, we might direct horizontally polarized photons to one detector, and vertically polarized photons to a different detector: each analyzer requires two detectors, as well as a device that separates photons according to polarization.

3. N. Gisin (2014), *Quantum Chance: Nonlocality, Teleportation and Other Quantum Marvels* (Springer International Publishing). I was also influenced by N. David Mermin's papers cited later in this chapter.

4. A. Zeilinger (2010), *Dance of the Photons* (Farrar, Straus and Giroux), based on E. P. Wigner (1970), "On hidden variables and quantum mechanical probabilities," *American Journal of Physics* 38: 1005–1009; and B. d'Espagnat (1995), *Veiled Reality: An Analysis of Present-Day Quantum Mechanical Concepts* (Addison-Wesley).

5. Assuming that the angle between polarizers is between 0° and 90°.

6. More precisely, we know that the other photon will pass through a polarizer set to the same angle as the first polarizer.

7. R. Penrose (2004), *The Road to Reality: A Complete Guide to the Laws of the Universe* (Alfred A. Knopf).

8. T. Maudlin (2002), *Quantum Non-Locality and Relativity*, 2nd ed. (Blackwell Publishing). My version of Maudlin's example appeared previously as the afterword to J. Brody (2017), "Hidden Variables," in M. Brotherton (ed.), *Science Fiction by Scientists* (Springer), 67–79.

9. N. D. Mermin (1994), "Quantum mysteries refined," *American Journal of Physics* 62: 880–887.

10. N. D. Mermin (1990), "Quantum mysteries revisited," *American Journal of Physics* 58: 731–734. I draw also from J.-W. Pan, D. Bouwmeester, M. Daniell, H. Weinfurter, and A. Zeilinger (2000), "Experimental test of quantum non-locality in three-photon Greenberger-Horne-Zeilinger entanglement," *Nature* 403: 515–519.

Chapter 5

1. They'll also disagree about the wavelength of light (Doppler effect), but this can be inferred from their disagreements about lengths and time intervals. Based on these disagreements, they'll also disagree about the speed of something that is moving relative to both of them.

2. Earth, of course, is rotating about its axis and revolving around the sun. The sun itself is moving in a complicated way around the center of the galaxy. Compared with the accelerating bus, however, we may imagine that Earth is at rest.

3. There are alternative explanations for the twin paradox that do not explicitly depend on acceleration (https://en.wikipedia.org/wiki/Twin_paradox).

Chapter 6

1. B. Skyrms (1982), "Counterfactual definiteness and local causation," *Philosophy of Science* 49: 43–50.

2. S. Hossenfelder (2014), "Testing superdeterministic conspiracy," *Journal of Physics: Conference Series* 504: 012018.

3. B. S. DeWitt and N. Graham (eds.) (2015), *The Many Worlds Interpretation of Quantum Mechanics* (Princeton University Press).

4. M. Schlosshauer (2005), "Decoherence, the measurement problem, and interpretations of quantum mechanics," *Reviews of Modern Physics* 76: 1267.

5. The "throwing out of the equation" process is called the collapse of the wavefunction.

6. A. Aspect, J. Dalibard, and G. Roger (1982), "Experimental test of Bell's inequalities using time-varying analyzers," *Physical Review Letters* 49: 1804–1807.

7. Three groups achieved this in 2015: B. Hensen et al. (2015), "Loophole-free Bell inequality violation using electron spins separated by 1.3 kilometers," *Nature* 526: 682–686; M. Giustina (2015), "Significant-loophole-free test of Bell's theorem with entangled photons," *Physical Review Letters* 115: 250401; L. K. Shalm et al. (2015), "Strong loophole-free test of local realism," *Physical Review Letters* 115: 250402.

8. A. S. Friedman et al. (2019), "Relaxed Bell inequalities with arbitrary measurement dependence for each observer," *Physical Review A* 99: 012121.

9. J. Handsteiner et al. (2017), "Cosmic Bell test: Measurement settings from Milky Way Stars," *Physical Review Letters* 118: 060401.

10. C. Abellán et al. (2018), "Challenging local realism with human choices," *Nature* 557: 212–216.

11. This video game is still available: https://museum.thebigbelltest.org/quest/.

12. J. Maldacena and L. Susskind (2013), "Cool horizons for entangled black holes," *Fortschritte der Physik* 61: 781–811.

13. A. Einstein and N. Rosen (1935), "The particle problem in the general theory of relativity," *Physical Review* 48: 73–77.

14. N. Gisin (2014), *Quantum Chance: Nonlocality, Teleportation and Other Quantum Marvels* (Springer International Publishing).

15. As in E. A. Abbott (1884), *Flatland* (Seeley & Co.).

16. Gisin, *Quantum Chance*.

17. T. Maudlin (2002), *Quantum Non-Locality and Relativity*, 2nd ed. (Blackwell Publishing).

18. A. J. Leggett (2003), "Nonlocal hidden-variable theories and quantum mechanics: An incompatibility theorem," *Foundations of Physics* 33: 1469–1493.

19. J. Cartwright (2007), "Quantum physics says goodbye to reality," *Physics World*, https://physicsworld.com/a/quantum-physics-says-goodbye-to-reality/.

20. O. Ulfbeck and A. Bohr (2001), "Genuine fortuitousness. Where did that click come from?" *Foundations of Physics* 31: 757–774.

21. C. A. Fuchs, N. D. Mermin, and R. Schack (2014), "An introduction to QBism with an application to the locality of quantum mechanics," *American Journal of Physics* 82: 749–754.

FURTHER READING

Ball, Philip. 2018. *Beyond Weird: Why Everything You Thought You Knew about Quantum Physics Is Different*. Chicago: University of Chicago Press.

Becker, Adam. 2018. *What Is Real? The Unfinished Quest for the Meaning of Quantum Physics*. New York: Basic Books.

Brody, Jed. 2017. "Hidden Variables." In *Science Fiction by Scientists*, ed. Michael Brotherton, 67–79. Cham, Switzerland: Springer International Publishing.

Bub, Tanya, and Jeffrey Bub. 2018. *Totally Random: Why Nobody Understands Quantum Mechanics*. Princeton: Princeton University Press.

Capra, Fritjof. [1975] 2013. *The Tao of Physics: An Exploration of the Parallels between Modern Physics and Eastern Mysticism*, 5th ed. Boulder: Shambhala Publications.

Fuchs, Christopher A., N. David Mermin, and Rudiger Schack. 2014. "An introduction to QBism with an application to the locality of quantum mechanics." *American Journal of Physics* 82: 749–754.

Gilder, Louisa. 2008. *The Age of Entanglement: When Quantum Physics Was Reborn*. New York: Alfred A. Knopf.

Gisin, Nicolas. 2014. *Quantum Chance: Nonlocality, Teleportation and Other Quantum Marvels*. New York: Springer International Publishing.

Greenstein, George S. 2019. *Quantum Strangeness: Wrestling with Bell's Theorem and the Ultimate Nature of Reality*. Cambridge, MA: MIT Press.

Herbert, Nick. 1985. *Quantum Reality: Beyond the New Physics*. New York: Anchor Books.

Kaiser, David. 2011. *How the Hippies Saved Physics*. New York: W. W. Norton & Company.

Kwiat, Paul G., and Lucien Hardy. 2000. "The mystery of the quantum cakes." *American Journal of Physics* 68: 33–36.

Maudlin, Tim. [1994] 2002. *Quantum Non-Locality and Relativity*, 2nd ed. Malden, MA: Blackwell Publishing.

Mermin, N. David. 1981. "Bringing home the atomic world: Quantum mysteries for anybody." *American Journal of Physics* 49: 940–943.

Mermin, N. David. 1985. "Is the moon there when nobody looks? Reality and the quantum theory." *Physics Today* 38: 38–47.

Mermin, N. David. 1990. "Quantum mysteries revisited." *American Journal of Physics* 58: 731–734.

Mermin, N. David. 1994. "Quantum mysteries refined." *American Journal of Physics* 62: 880–887.

Scarani, Valerio. 2006. *Quantum Physics—A First Encounter: Interference, Entanglement, and Reality*. Trans. Rachel Thew. New York: Oxford University Press.

Siegfried, Tom. Jan. 27, 2016. "Entanglement is spooky, but not action at a distance," *Science News*.

Zeilinger, Anton. 2010. *Dance of the Photons*. New York: Farrar, Straus and Giroux.

INDEX

The MIT Press Essential Knowledge Series

JED BRODY is Senior Lecturer in Physics at Emory University. He has written two science-fiction novels, *The Philodendrist Heresy* and *The Entropy Heresy*, originally published by Moon Willow Press and republished in 2019 by Stormbird Press.